CARE OF CREATION

D1219952

CARE OF CREATION

Christian Voices on God, Humanity, and the Environment

Edited by Joseph Coleson

wesleyan
publishing
house

Indianapolis, Indiana

This volume is dedicated to the memory of John and Charles Wesley, William Wilberforce, Orange Scott, Luther Lee, William and Catherine Booth; to J. A. L. and World Hope International; and to the legions of like-minded persons serving God, God's people, and God's good earth, simply because God calls and they have heard.

CONTENTS

Part 3: Care of the Environment

ACKNOWLEDGMENTS

Wesleyan Theological Perspectives is made possible through the support of the Wesleyan Education Council, comprising the presidents and deans of four liberal arts universities—Houghton College, Indiana Wesleyan University (IWU), Oklahoma Wesleyan University (OWU), and Southern Wesleyan University (SWU)—and of Bethany Bible College. These are the educational institutions sponsored by the North American General Conference of The Wesleyan Church (TWC). We are grateful for their continued generous support of this and other projects of our steering committee, comprising the university chairs in religion, the dean of the Bible college, and the series editor.

This fifth volume in the series was planned and shaped by Kelvin Friebel (Houghton), Kenneth Gavel (Bethany), Kerry Kind (TWC), Steve Lennox (IWU), Roger McKenzie (SWU), Mark Weeter (OWU), Joseph Coleson (Nazarene Theological Seminary, chair), and Charlotte

Coleson (assisting as recording secretary) in a focused set of meetings in February 2008, in Indianapolis, Indiana. We thank all of them for their invaluable contributions.

Kerry Kind, General Director, Department of Education and the Ministry, The Wesleyan Church; Don Cady, General Publisher, Wesleyan Publishing House; and Craig Bubeck, editorial director, Wesleyan Publishing House have been unfailingly supportive of this project and wise in their counsel from the beginning. Kevin Scott, editor, Wesleyan Publishing House, provided significant in-house editorial support.

The board of General Superintendents has given their enthusiastic endorsement of this project from its inception; we thank all its members, past and present. They have led the way in the acceptance and use of these volumes across our beloved Church.

You, our esteemed readers, have read and referred to the volumes in this series as they have become available; you have used them with and recommended them to an ever-growing circle of parishioners, students, and other friends. More than we can express, you have our heartfelt thanks.

As always, we gladly render the glory that belongs solely to God.

INTRODUCTION

Joseph Coleson

A s I wrap up my editing work on this volume, a 7.0-magnitude earthquake has devastated Port-au-Prince, Haiti, and the surrounding area. As of yet, no good estimate of the final death toll is available, but officials are saying it could go as high as two hundred thousand persons killed.

Within two days of the earthquake, two "explanations" made the news. First, a well-known Christian television personality blamed the earthquake on a pact with the devil that Haiti's founders made during their revolution to throw off French colonial control at the turn of the nineteenth century. Then, a well-known actor (not known as a Christian) said it was the result of the failure of the Copenhagen conference in December 2009—barely a month before the earthquake—to reach a meaningful accord on dealing with climate change. Ill-informed and illogical, both attempts at explanation illustrate the breadth and depth of misunderstanding across our culture of God's creation upon this earth and of God's creation stewardship mandate upon us.

That divine mandate—to care for God's good creation on this earth—is the subject of this book. Genesis 1 and 2 are clear about this as God's creation intention; human beings are God's stewards, responsible for the care and protection of the earth and all its creatures.

This responsibility extends both to our fellow human beings and to the rest of the creation on this earth—the earth itself, its plants, its animals, and all its interrelationships. We call these by various titles among ourselves such as, cultures, societies, nations, ethnic groups, tribes, clans, or families. With respect to the rest of the living creation, we call it, most simply, the biosphere—the sum total of all the earth's ecosystems.

Our responsibilities to each other are different in a number of respects from our responsibilities to the rest of creation. In God's creation intentions, we were not given stewardship dominion over each other, as we were over the earth and the rest of its creatures. However, our early and continuing rebellion against God complicates matters; we deal with that issue at various points in this book.

As one result of that first rebellion, the great catastrophe many theologians call the fall, we became much less inclined to care for creation as God intends and also much less adept at it. However, God has not excused us. In the final reckoning, we will be asked for an accounting of our stewardship, and we will not be allowed to pass on the question. We have not cared for each other appropriately, either, and that, too, will be part of the final accounting.

We ought not to approach these issues from a negative perspective, however. God has built the joys of relationship and stewardship care into our DNA, as it were. Even in our estrangement from God and from each other, one of the greatest of human joys is experiencing genuine relationship with each other in the many patterns appropriate relationships can display when we are careful to weave them in love and integrity. Even in our estrangement from the earth, another deep and abiding joy is nurturing and watching the growth of living things in both of God's good biological realms on this earth—the plant kingdom and the animal kingdom.

Part 1 of this book, the first four chapters, lays out the biblical theology of care for creation beginning with God's original creation design as reported in Genesis 1 and 2. Let me urge you not to skip these chapters. They are both relevant and critical in their laying of a solid foundation for the case for all Christians' passionate involvement in creation care for the right reasons.

In part 2, chapter 5 is an introduction to issues raised by the amazing advances in our understanding of human genetics in recent years. Chapter 6, under the broad headings of abortion and euthanasia, deals with ethical issues surrounding the beginning and the ending of life. Chapter 7 highlights some of the ways we have found, in our sinful greed, to exploit our brothers and sisters also created in God's image just because we can. More importantly, Jo Anne Lyon challenges us to work toward the elimination of these gross injustices as our forebears in the Methodist/Wesleyan/Holiness tradition did in their generations.

In its four chapters, part 3 introduces four important sets of issues involved in taking our creation mandate seriously with respect to the earth and its non-human living populations. Chapter 12, then, comprises a call to action from Matthew and Nancy Sleeth, who recently have emerged at the forefront of evangelical Christian activism in creation care issues, particularly with respect to Christian living in and for the environment God has charged us with keeping. They have been living creation care for years now; their story and advice should inspire you.

Together with all the contributors, I pray this book will give you joy at many places. We also pray it will open your eyes and challenge you, in ways you have not been challenged previously, to embark upon, renew, or intensify your efforts to fulfill our creation mandate toward God's good earth and its inhabitants, and to minister in healing ways to our sisters and brothers who need the love and care we can extend. Even a cup of cold water given in Jesus' name is precious and eternally consequential. God's *shalom* to you as you read and act upon God's promptings, both individually and with those you count as your local family of faith.

PART 1

CREATION, ALIENATION, REDEMPTION

GOD'S CREATION MANDATE FOR HUMANITY

Joseph Coleson

So God took the 'adam and settled him there in the garden of Eden,
to serve it and to keep it. . . .
This one, this time, is bone of my bones, and flesh of my flesh.
She shall be called woman, for out of man was this one taken.

—Genesis 2:15, 23 (author's translation)

Here in my home office, I keep four hats on top of the bookcase to the left of my desk. The interesting thing about that is I seldom wear a hat, even in winter—but here they are.

When we say of someone, "She wears many hats," we usually mean that she fills a variety of roles in her job, church, family, or elsewhere.

> *A Heaven on Earth: for blissful Paradise*
> *Of God the Garden was, by him in the East*
> *Of Eden planted.*
>
> —John Milton

As crown of earthly creation, made in God's image, all humans wear several hats. God has placed us here for many good reasons—all of them to our good if we will but trust that God is good.

The Story of Creation

The purposes of this opening chapter are to discuss: (1) what the biblical creation story teaches about how we are to treat each other and (2) what it teaches about how we are to treat the earth and its non-human inhabitants.

We will focus primarily on Genesis 1–2 for two main reasons. First, the Bible begins with creation, and beginnings are always important. Second, the narrative of Genesis 1–2 is the only wide-angle picture we have of God's intentions for creation before the distortions caused by human rebellion.

Christians take the Bible seriously as God's special revelation. The Bible tells the story that natural revelation can only begin to tell, though nature strongly suggests it, as Psalm 19:3 reminds us, "There is no speech, and there are no words; their voice is not heard" (author's translation). So what God revealed in words to ancient Israel invites and requires our close, careful attention to ensure that what we read and hear is what it actually says.

"Close, careful attention" can get a bit technical at times, but bear with me. Technical doesn't have to mean boring, despite the common assumptions in our culture. These first chapters of Genesis matter; they will reward our work with them.

To Love and Cleave

We begin with care for humans for several reasons: (1) in Genesis 1, the author asserts humanity's unique place in the world before recording God's mandate for us to care for the earth; (2) we need a biblical understanding of our place in creation in order to fulfill the creation care mandate appropriately and to know when we have done so or failed to do so; and (3) we need to be firmly rooted in a biblical theology of creation so that care for the earth and its non-human inhabitants will not devolve into worship of creation or neo-paganism, rather than stewardship dominion.

Days of Creation

In a narrative so beautiful it often is called exalted or poetic prose, Genesis 1 presents a summary of God's good creation. From the initial statement, "In the beginning God created the heavens and the earth," it moves immediately to its main subject, "Now the earth was . . ." (Gen 1:1–2).

Genesis 1 presents God's creation upon the earth as a series of three pairings. "Let there be light . . ." (first day, vv. 3–5) is paired with the making and placing of the light-bearers (fourth day, vv. 14–19). The creation of the expanse dividing the waters above and below (second day, vv. 6–8) is paired with the population of the waters below and the expanse above by the creatures of the waters and the air (fifth day, vv. 20–23). The gathering of the waters below so the dry land could appear (third day, vv. 9–13) is paired with the creation of land creatures, including humans (sixth day, vv. 24–31). The real culmination of this narrative's creation week, and of its theology, is the seventh day of rest, the first Sabbath (2:1–4), but that is a discussion for another occasion.

Creation of the *'Adam*

God's work in the Genesis 1 narrative climaxes in the creation of humans. Hebrew *'adam* (Hebrew "human being"; v. 26) is a collective noun, which means it may designate either an individual or a group, depending on its context. Moreover, *'adam* does not mean man or male only, but rather human(s), as Genesis 1:27 will make clear. In case we should miss it, though, it is even more emphatic in the summary of Genesis 5:2, "Male and female [God] created them and [God] blessed them, and [God] called *their* name *'adam* in the day of their creation" (author's translation).

Genesis 1:26 says, "Then God said, 'Let us make *'adam* in our image, according to our likeness" (author's translation). As God's creative acts upon the earth reach a climax, the importance of this particular creation is signaled by God's pausing to hold a heavenly council. Did God confer within the Trinity? With already-created angels? Perhaps both; we cannot rule out either. What we do know is no other act or work of creation is accorded this honor and significance. This honor is given only to the 'adam.

The decision made, the narrator records its enactment: "So God created ['adam] in [God's] own image, in the image of God he created him [the pronoun here also must be a collective, or the next clause makes no sense]; male and female [God] created them" (v. 27). It begins here, at the introduction of the human species, and it continues throughout these two chapters until the human, turning away, mars the whole. What begins, you ask? The insistence of the text that both male and female are human, that both are created in God's image, and that both share in God's blessing and mandate.

God's very first communication to the newly formed humans is a blessing, but a blessing in the form of a five-fold instruction. All five verbs are imperative (in command form) and all five are plural, "God *blessed* them and said to them, '[You] be fruitful and [you] increase in number; [you] fill the earth and [you] subdue it. [You] rule over . . .'" (v. 28, emphasis added). Both fertility and stewardship are God's blessing, but also God's mandate. God gave both fertility and stewardship to the woman and to the man, jointly and equally. Whether every human would have produced offspring had not sin intervened, we cannot say. We do know God freely bestowed this blessing as a vital part of the natural order. We need not beg, bribe, or cajole fickle nature gods for it, as the ancient pagans thought.

This free and abundant blessing, bestowed at the inception of human life, coheres with the later scriptural portrait of God as a "compassionate and gracious God, slow to anger, abounding in love and faithfulness" (Ex. 34:6). God intended the first humans to practice being, living, and loving in the image of God, and to teach their children to do likewise. It did not happen that way, but this was God's intention.

From Solitude to Union

Genesis 2 does not contradict, but complements Genesis 1. It provides detail in the picture of human creation that would destroy the literary quality of Genesis 1 if forced into its tightly-woven narrative. Genesis 2:18–25 records the second and final step in the creation of the human species; God's making of the woman.

First, we must not miss the narrator's extreme care to show that the woman was human, along with the man. Verse 18 records the only "not good" of the creation narratives, "Then God said, 'It is not good for the *'adam* to be alone. I will make for him an *'ezer cenegdo'* [A-zer ceh-neg-DOE]" (author's translation). We cannot detail the Hebrew evidence here, but (anticipating the gender distinction) we should translate this phrase, "a power like him, facing him as equal." In this context, "a power" is another human. "Like him" means "of the same kind or species"; the animals the *'adam* would name were not of his species. "Facing him as equal," a preposition with an attached pronoun, repeats both ideas, "a power" and "like him," but in different words; repetition for emphasis is an important feature of the biblical text.

God then caused the larger land and flying creatures to present themselves to the *'adam* for naming (v. 19). This served two purposes. First, the *'adam* began to exercise the stewardship dominion God would mandate shortly. Second, the *'adam* soon realized none of these creatures was the power like him of which God had spoken. The *'adam* was ready for the second act of human creation.

Taking from the side (not the rib only) of the sleeping *'adam*, God constructed another (v. 22). The man recognized her as human the moment he saw her. "This is now bone of my bones and flesh of my flesh; she shall be called 'woman,' for she was taken out of man" (v. 23). He saw she was like him, of the same kind and species, a power who could face him as equal.

The narrator's own editorial instruction on this, given in and to a distinctly patriarchal culture, is enough to remove all doubt, were there any. "For this reason a man will leave his father and mother and be united to his wife, and they will become one flesh" (v. 24). Mature adults leave the hierarchical authority that nurtured them to effective adulthood, and establish a new partnership based on mutuality, equality, parity, and reciprocal regard. This is God's creation intention.

We must skip over most of Genesis 3, but even in this chapter reporting the great disaster, we find affirmations of God's original perfect intentions. One of these is the notice of God's habit of communing with the human pair in the cool of the afternoon (v. 8)—in the Middle East, a refreshing time of day, when the midday heat diminishes. Here, vocabulary, grammar, and syntax combine to show God's own delight in the creation and intimate fellowship with the human pair, even his disappointment when, one day, they failed to show.

God cursed the serpent for its role in the human rebellion (v. 14), but did not curse either the woman or the man. God predicted that the man henceforth would dominate the woman (v. 16). However, even apart from the previous narrative, it is clear from this verse alone that this was not God's creation intention or mandate. It is not even God's arbitrary sentence, as though several were available, and God chose this one for the woman. Rather, man's domination of woman is one of many disastrous but natural consequences of sin, as their mutuality of interdependence was replaced by greed for dominance. I repeat: man's domination over woman is not God's creation intention.

It is—first, last, always, and everywhere—the result of sin. That Genesis 3 teaches and illustrates this so vividly confirms that our reading of Genesis 1 and 2 is correct.

Caring for One Another

Again, if space permitted we could develop several implications. We must mention only three briefly, and move on.

First, every human is and reflects the image of God. This image was distorted, but not destroyed, by the action of our first parents. Thus, every human is of measureless worth. Woman as well as man, man as well as woman, is created by the finger of God, in the image of God, and gifted with the breath of God. By extension, no other reckoning can make any human into less than the image of God: ethnicity cannot; language cannot; citizenship cannot; education or lack thereof cannot; wealth or lack thereof cannot; and pedigree cannot.

Any theology, philosophy, custom, or action that violates or devalues this truth is not from God but from another source entirely. Another way of saying this is that God has a place at the table for every person, an equal place in the family of God. God expects us to affirm and practice this ethos today, both for its own sake and as preparation for the days to come.

Second, God intended marriage as the daily setting for human love and care for each other in its most intimate expression. God intended family as the setting where children are loved and cared for completely. Family life is designed to teach that love does not hoard power, nor use power abusively; love empowers. God designed the family, as a society in miniature, to be a model, a laboratory producing adult powers able and willing to become godly, loving members of the larger society beyond our kin, but within our ken.

Third, permanent or oppressive hierarchy is a result of sin. It is not part of God's creation purposes; rather, it is antithetical to them.

Relationships of health and integrity flourish not because of, but in spite of, a legitimate hierarchical atmosphere, however benign it may appear on the surface or from a distance. Any and every legitimate hierarchy is formed for a limited purpose, for a limited time, with carefully limited powers, and subject to rigorous accountability. This includes parents over their children, teachers over their students, pastors over their congregations, managers over their employees, and governments over their citizens. It includes every hierarchical arrangement because, in this world drunk with power and in pursuit of power, it is true more than ever that power corrupts, and absolute power corrupts absolutely.

To Serve and Keep

The second purpose of this chapter is to focus on the role God assigned us—already a very important "hat" at our creation—to serve and to keep the earth as stewards, as God's regents, responsible under God for the well-being of the earth and all its creatures, not just for ourselves and our own good. Indeed, the creation theology of the Bible teaches that our own good cannot be sustained apart from the good of the rest of creation we were appointed to serve and to keep.

Of course, we bungle this assignment; our first parents began the bungling with their fateful decision to turn from God, as told in Genesis 3. But their tragic decision did not nullify our charge to serve and to keep God's creation here on this earth, though it made the task more difficult. If we are to wear this hat well, we will begin by discovering what God intended from the beginning.

The Stewardship Mandate

Following the report of God's creation of the 'adam (Gen. 1:27), Genesis 1:28 records the stewardship mandate. This fact is important in and of itself. The very first words of God to the human pair are this combination blessing and mandate to exercise dominion over the rest of God's earthly creation, including even the earth itself! The entire verse reads, "God blessed them; and God said to them, 'Be fruitful and multiply, and fill the earth, and subdue it; and rule over the fish of the sea and over the birds of the sky and over every living thing that moves on the earth'" (NASB).

It is also significant that this blessing is presented as an imperative, a series of five plural verbs of command: [you all] be fruitful; [you all] multiply; [you all] fill; [you all] subdue; [you all] rule. We have already noted these plural forms as evidence that God intended both male and female to be God's agents upon the earth. Also important is the identification of blessing with instruction. Far from restricting human freedom and happiness, God's commands or instructions are for our good and for the good of all the earth. How would we acquire knowledge, experience, wisdom, and all else that makes for *shalom* (Hebrew "peace, wholeness, total well-being"), unless God had, in the infancy of our kind, started us on the right path with instruction? How would we do so, unless God had continued, even after our rebellion, to teach and counsel us through Scripture and other complementary means?

Finally, it bears repeating that God did not abolish or modify these instructions when we turned from God and became incapable of fulfilling them as originally intended. We still are responsible. As Christians, we stand in awe of, but also take delight in, God's gracious invitation to partner in working toward the fullness, the *telos* (Greek "end"), of the restoration Christ effected by his death and resurrection.

Agents of God's Care

At several points, Genesis 2 confirms that God created humans for responsible stewardship upon the earth. Verse 5 notes "no plant of the field had yet sprung up, for the LORD God had not sent rain on the earth and there was no ['*adam*] to work [serve] the ground." Wheat, barley, and other cultivated grains are meant here. It is not that these species did not exist, but they need special conditions to produce enough crop to be useful as a source for food. In Canaan, they needed the rain provided by God, but also the human work of the soil— plowing, planting, tilling, and harvesting. The earth does best in the unimpeded exercise of this divine-human partnership.

Genesis 2:7 reports that God formed the first '*adam* from the '*adamah* (Hebrew "earth"). We may translate this as "earthling from the earth" or "human from the humus." The word play is intentional in the Hebrew text. Our physical origin from and connection to the earth is an intrinsic reason for humans to care about creation.

Genesis 2:8 says, "God planted a garden eastward in Eden; and there he put the ['*adam*] whom he had formed" (KJV). After a fairly lengthy description of the garden, the author resumes the narrative of human creation. Verse 15 records two reasons for placing the human in the garden: to work or serve it, and to keep, watch, guard, and pro- tect it.

This verse demonstrates that the common idea that work is a curse, the result of sin, is false. Far from being a curse, work is a blessing and one of the ways we function in the image of God. God worked in the creation of the universe, including this earth. Then God invited the first human to partner in the work of maintaining and pro- tecting the garden of Eden, a kind of microcosm of the whole earth, a place to learn what work means and how to do it.

The verbs used in Genesis 2:8 help us understand the meaning of the earlier verbs (1:26, 28) "rule over" and "subdue." God did not make humans the agents of God's care over the earth and its creatures to misuse and abuse them. Rather, we are charged with serving— with working, sometimes long hours or in difficult or even dangerous circumstances, to ensure by our physical and mental labor the well-being of the earth, its vegetation, and its creatures. Moreover, we are charged with keeping them—guarding them, watching over them, and protecting them from harm. Examples are shepherds and ranchers who care for their stock in harsh weather before seeking their own shelter; more than a few have died saving their animals.

We should not assume it was unnecessary to protect creation from harm in the garden paradise. Catastrophic harm actually did come upon all creation due to the first humans' blatant disobedience. If it could happen then, how much more ought we now to be vigilant for the well-being and protection of all God has charged us with?

It is important to note that (together with many other implications) the "one flesh" of Genesis 2:24 means man and woman share equally in our God-appointed care for the earth and our fellow-creatures.

Conclusion

We have pondered, all too briefly, what we may learn from Genesis 1–2, the only extended narrative we have of God's creation intentions unmarred by humans turning away. To fudge a bit and peek over into Genesis 3, we should note God's great delight in experiencing creation with God's creatures and the disappointment in God's question, "Where are you?" (vv. 8–9; the Hebrew is more vivid than the translations can render). This alone is reason enough for us to care passionately about serving and keeping. God delights in experiencing creation with us!

Of course, we do not live amidst God's lavish provision in Eden anymore. God predicted and it came to pass, the earth's diminished capacity to yield its bounty to our hand. The work of serving and keeping still occupies us, but now it is more troublesome and less rewarding, even with the great advances of the last two centuries. At life's end, we return to the earth from which we were taken.

But all is not lost! In what J. R. R. Tolkien called the great *eucatastrophe*, God in Christ has reversed the great disaster. God promises the earth and our fellow creatures will share with us in the eschatological renewal. We cannot know when it will be revealed in all its fullness. We are invited, though, to practice wearing the hat—to live as though it has begun, because it has begun. As you read through this book, look for ways to serve and keep the earth, and to respect, honor, and love others as Genesis 1 and 2 teach that God created and calls us to do.

Suggestions for Reflection and Action

1. Chapter by chapter through this book, ask yourself, "What might this world have been now, had our first parents not turned from God? What can I, and we, do now to practice and demonstrate God's already-yet-still-to-come restoration of all things?"

2. For seven days, record your interactions with others unlike yourself by reasons of gender, ethnicity, economic status, health, or other markers. Ponder whether and how you have treated each as a brother or sister also created in, and exhibiting, the image of God. If you have not done so, how do you need to change?

3. Find a way to (re)connect with the earth, the soil, the trees or other plants, or animals large or small. Plant and tend a garden, however small; it will help you (re)connect with God's blessing and mandate.

For Further Reading

Holy Bible, New Living Translation, or another easy-to-read version. Genesis 1–2 outlines God's creation upon this earth; Genesis 3, the turning away of our first parents. The rest of the Bible is about God's redemption and restoration of all creation in Christ. Reading the Bible with this in mind will give you the big picture as you've never seen it before.

Coleson, Joseph. *'Ezer Cenegdo: A Power Like Him, Facing Him as Equal.* Wesleyan/Holiness Women Clergy, 1996.

It is in need of revision and expansion, but this booklet details much of the evidence for God's creation intention of gender (and other) equality summarized in the first section of this chapter.

Fretheim, Terence E. *God and World in the Old Testament.* Nashville: Abingdon Press, 2005.

This volume is an advanced seminar on, as its subtitle puts it, "A Relational Theology of Creation." It is not for everyone, but those who persevere will find themselves reading a theology of creation and restoration that sounds Wesleyan on many points, though Fretheim is a prominent Lutheran Old Testament scholar.

Walton, John H. *The Lost World of Genesis One.* Downers Grove, Ill.: InterVarsity Press, 2009.

The best treatment of Genesis 1 that I've come across in a long time, and I read everything I can find on Genesis 1! Walton's basic thesis is that function (rather than material "existence") is the key to the chapter's presentation of God's creation work on this earth. This work is a model of clear, concise writing by an academic for a non-specialist audience. You will not get lost in this book, you will enjoy it, and it will open Genesis 1 to you in new, inspiring, and even joyful ways.

THE DISASTROUS RESULTS OF HUMANITY'S FALL

Kelvin G. Friebel

Cursed is the ground because of you.

—Genesis 3:17

The debate undoubtedly will continue over whether humans are directly responsible for global warming. But Christians acknowledge two realities. First, the world is not the paradise of Genesis 2 as God originally made it. The process of de-creation disrupts and undoes the original intent

> The earth is on the verge of ecological collapse, and we are the cause.
>
> —Richard Cizik

of creation and is an ever-present part of our environment. Second, this marring of creation stems from human sin.

Disruption and De-Creation

Genesis 1–2 gives the reader a panoramic overview of God's creation intentions (see previous chapter). Genesis 3 reports the disruptive event as the great disaster.

Distortion of Nature

God's original work on this planet (Gen. 1–2) produced a creation functioning as God designed it. Although the creation we presently inhabit still operates according to an infrastructure of natural laws, we recognize that nature has been radically distorted from God's original creative intent of a harmonious interdependence. Nature encompasses more than beautiful sunsets, breathtaking landscapes, majestic snowcapped mountains, and amber waves of grain. She also exhibits predatory animals, hurricanes, tornadoes, volcanic eruptions, forest fires, earthquakes, droughts, and floods.

Distortion of Relationships

What has brought about this distortion within creation? According to Genesis 3, the primary factor was Adam and Eve's sin. As a consequence of their deliberate act of disobedience against the explicit divine command (Gen. 2:17), not only did the judgment of death come upon humans (3:19), but their act also caused the ongoing disruption of all the harmonious interconnected relationships which were part of creation as portrayed in Genesis 2.

First, sin disrupted the relationship between humans and God. This is illustrated by Adam and Eve hiding themselves from the presence of the Lord, and their fear of interacting with him (3:8, 10).

Second, sin disrupted the relationship between man and woman. Rather than partnership and connectedness (2:18, 23), the relationship between man and woman would now be characterized by each seeking to dominate the other (3:16).

Third, sin disrupted the relationship between humans and the inanimate creation. Originally there was a mutual benefiting, as the garden produced food for humans (2:9, 16), and the humans cultivated (served) and cared for (kept) the garden (2:15), facilitating its flourishing. But due to Adam and Eve's sin, God placed a curse on the ground (3:17). No longer would it respond the same way to the cultivating efforts of humans; rather, it would "produce thorns and thistles" (3:18). Though still providing food for the humans (3:18), food now would come only at the expense of hard toil on their part (3:17, 19).

Finally, sin disrupted the relationship between humans and the rest of the animate creation as well. In the garden, the relationship between animals and humans was non-aggressive and companionable (2:18–20). This conclusion is confirmed also by prophetic portrayals of the future paradise based on Eden imagery; two examples are Isaiah 11:6–9 and Amos 9:13–15. This alteration of the relationship between humans and the animal kingdom is also highlighted in God's blessing given to Noah and his family after the flood. It emphasizes that animals would live in fear of humans because now animals, too, would be food for humans (Gen. 9:2–3).

Collateral Damage

In summarizing the effect of human moral failure upon nature, Genesis 3 emphasizes the symbiotic interconnectedness of all creation. A disruption in one part of creation affects the whole. Human violations in the moral realm, though perpetrated against God or against fellow humans, often cause collateral damage to the nonhuman creation. Genesis 3 reports that even though the humans did not sin against the ground nor did the ground do anything wrong in the roles God had created it for, the ground, an innocent bystander, was cursed, introducing a permanent de-creational element into nature. Paul characterized this as the creation being "subjected to frustration" and "bondage to decay," even as "groaning" to be released (Rom. 8:20–22). Creation is in such a condition, not by its own action, choice, or volition, but because of what humans have done against God.

Tainted but Still Useful

The events of Genesis 4 demonstrate that though creation is altered from its original design, the effects of human sin have not entirely blotted out its original goodness. Rather, evil now coexists alongside nature's intrinsic goodness. Humans learned to utilize the resources of creation for beneficial developments, such as building cities and making musical instruments (Gen. 4:17, 21). These creative advances in civilization and the arts were tied to advances in technology as humans used bronze and iron to make tools (4:22). Even as humans became more sinful, they still retained the creative skills and ingenuity to produce good things from the good resources of the cursed earth.

Human Responsibility for De-Creation

Genesis 3 teaches that the foundational cause of the distortions in creation is human sin. However, when an ecological disaster occurs, the question usually arises of how to explain that specific event theologically. Are all ecological catastrophes directly caused by human sin?

With respect to specific incidents, we should understand the connection between human sin and the disruption of creation from one (or sometimes more than one) of four perspectives: (1) the natural condition of the fallen world; (2) the long-term effects of unwitting human behavior; (3) human abuse of nature; and (4) acts of divine judgment resulting in ecological disaster. The first three we will discuss in this section; the fourth is prominent enough in Scripture to be treated separately.

A Fallen Creation

While God is constantly and intimately involved in the ongoing functioning and sustaining of creation, God also created nature to operate according to natural laws. These laws were affected to varying degrees by the de-creational elements introduced by Adam and Eve's sin, but God still lets creation function according to its intrinsic laws. Thus, some ecological disasters are merely the consequences of the natural world after the fall. They just happen as part of the very fabric of fallen creation.

The Bible recognizes these de-creational elements as inherent within the natural world as it now exists. Wilderness, predatory animals, and storms are part of the larger ecosystem. As one example, Deuteronomy 32:10 speaks of a desert or wilderness wasteland area using the same term as Genesis 1:2, "formless" (*tohu*). Its condition reflects that of the earth before God's creative acts of shaping and development.

Volcanic eruptions, hurricanes, tsunamis, forest fires ignited by lightning strikes—all these produce profound environmental devastation and ecological changes. Most are not attributable to lack of creation stewardship on the part of humans or to God's specific de-creational acts of judgment. They occur simply because nature is functioning as nature, given its altered condition due to Adam and Eve's sin.

Unintended Consequences

Some ecological difficulties are the consequences of long-term human behavior which, when initiated, was not intended to abuse nature. Because we have limited knowledge and cannot assuredly know future consequences of our actions, we do things which seem to be appropriate at the time and which arise from very good motives. Only in retrospect, then, do we learn we have unintentionally caused harm to our environment.

For example, only recently have we discovered that using asbestos in insulation and lead in house paint have serious medical implications. Likewise, the inventors of the gasoline engine and the automobile were not trying to pollute the air with emissions and deplete the earth's fossil fuel supply. Yet these have become unforeseen results of their vastly expanded use. When advancements in knowledge reveal that our actions are unhealthy for the environment but we fail to alter our behaviors, we are abdicating our mandate to be good stewards of the earth.

Intentional Abuse

Some ecological disasters result from our conscious, deliberate abuse of the environment and its resources. We attempt to satisfy our self-centered desires without regard for non-human creation or other segments of humanity. In our self-absorbed mindset, we do not seriously consider how our treatment of creation affects others, whether they are in other parts of the world or will belong to future generations. Examples from American history include the commercial hunting to extinction of the passenger pigeon, the near-extinction of the American bison (buffalo) and, until fairly recently, the reckless strip mining of coal in parts of Appalachia and beyond.

The Bible contains few direct statements or narratives of human abuse of creation; an exception is the ravages of war on the ecosystem. Invasion by an army often resulted in the land being laid waste, denuded not only of humans but of livestock and crops (see Lev. 26:33; Deut. 28:51; 2 Kings 19:23–24). When an enemy army invaded, it lived off the land, pillaging its resources for consumption or to use as implements of war.

The scarcity of examples of the abuse of nature is understandable, given the original social contexts in which the biblical revelation was given. Ancient Middle Eastern societies did not possess the means of abusing the environment that are prevalent in our urbanized, technologically advanced, consumer-driven, and globalized society. Neither overpopulation nor ratios of population to resources pushed the limits as commonly happens today. Unlike today, waste was negligible.

What little waste was generated was largely biodegradable—imagine a world without chemicals, without plastics, or without disposable wood, fabric, or metals of any kind. The only large-scale waste from sites within the biblical world is the abundance of potsherds (broken pieces of ceramic pottery) and their effects on the environment are as good as zero.

Within the economies of the biblical times, with rare exceptions, most food supplies and other natural resources were raised, harvested, or extracted from the local area or region. The vast majority of the population in biblical societies existed at a subsistence level. The technologies employed in farming, mining, and transportation would be called primitive by many today. Everything about production, transportation, and consumption in biblical times stands in sharp contrast with today's hyper-consumerist global economy that depletes natural resources at unprecedented rates. Ancient human abuses of creation could not have been of the same magnitude as today's abuses. Thus, the biblical authors did not need to address creation abuse as extensively as responsible stewardship calls us to today.

Divine De-Creation

The most common reason the Bible gives to explain specific ecological disasters is God's judgment upon humans for our disobedience. Frequently, as God acts in judgment, the rest of creation is also severely affected. Nature commits no crimes against God; nevertheless, nature becomes an object of divine judgment, an innocent victim. We have noted already the first biblical example of this in God's pronouncement of the consequences for the earth of Adam and Eve's sin (Gen. 3). Throughout Scripture, it is consistently evident in: (1) early biblical narratives describing acts of divine judgment, (2) the curses set forth in the law for disobedience, (3) the theology of judgment in the prophetic books, and (4) the apocalyptic literature of the book of Revelation. All these are rooted in a common theology of retribution, but their chronological span and the variety of their genres lends them added significance, beckoning us to closer attention.

In Biblical Narratives

We will discuss briefly three well-known stories as examples of biblical narrative reporting of God's judgment as de-creation. Two are from Genesis; one, from Exodus.

The Flood. Our first example is the Genesis flood. The first half of the narrative (Gen. 6–7), especially, portrays the flood as an act of de-creation, using specific terminology and broader imagery harking back to the creation account of Genesis 1. In verses 6:20 and 7:14, both the classes of animals listed and the phrase "according to their kinds" (7:14) draw upon Genesis 1:21, 24–25. As in 1:30, the animals are referred to as having the "breath of life" (6:17; 7:22). The command to Noah regarding food in 6:21 is based in God's provisions for food in 1:29–30. The sequence of seven days plays a significant role in the flood narrative (7:4, 10; 8:10, 12), as it does in the creation account.

With respect to the broader imagery, the separations of the creation week recorded in Genesis 1 are systematically undone in 7:11–24. First, "the springs [fountains] of the great deep burst forth, and the floodgates [windows] of the heavens were opened" (7:11), obliterating the separation of the waters above from the waters below on creation day two (1:6–7). The deluge continued until again the waters covered all the land (7:19–20), reversing God's separation of the terrestrial waters from the dry land on day three of creation (1:9–10). All living creatures on the earth perished in the flood (7:21–23), countering the divine activity of creation's day six (1:24–27). Thus, when one comes to the end of Genesis 7, the world looks as it did at the beginning of creation (1:2). The earth is covered in water (7:24). It is uninhabitable (formless) so, obviously, it is without inhabitants (empty or void).

Human wickedness was the reason for the massive destruction of the earth through the flood (6:5). But the judgment decreed in Genesis 6:7 was the blotting out not only of humans, but also of the larger land animals, the smaller creeping ones, and the birds; even the land vegetation was destroyed. In this case, while human sinfulness evoked God's judgment, the divinely executed judgment produced an ecological disaster of enormous proportions.

Sodom and Gomorrah. Our second narrative example is the destruction of Sodom and Gomorrah reported in Genesis 19. God reduced this well-watered area around the Dead Sea to a wilderness wasteland devoid of human, animal, and plant life (Gen. 19:25; see also Deut. 29:23; Jer. 50:40; Zeph. 2:9). This judgment was evoked by depraved human behavior (Gen. 18:20–21) and resulted in the complete devastation and alteration of the ecosystems of the Dead Sea region.

The Exodus Plagues. Our third example of God's judgment creating ecological disaster is the first nine of the ten plagues upon the Egyptians recorded in Exodus 7:14—10:29. Nine times the nation of Egypt experienced catastrophic environmental changes as God punished them for their cruel treatment of the Hebrews and Pharaoh's refusal to let God's people depart. Any one of these plagues by itself was an ecological disaster of major magnitude, to say nothing of the cumulative effect of all of them together. Egypt's natural environment was devastated, its agriculturally dependent economy ruined.

In all these examples, nature functioned in dual roles, both as recipient of the consequences of a judgment directed against humans and, at the same time, as God's instrument for inflicting judgment. In Genesis 3, the cursed ground became the means through which God brought hard labor to humans. In the flood narrative, both animate land and sky creatures and inanimate land vegetation were destroyed through the created rain and seas. At Sodom and Gomorrah, the land was devastated by God's rain of burning sulfur from the sky (19:24). In the plagues against Egypt, God used creatures (frogs, gnats, flies, locusts), meteorological elements (hail, lightning, darkness), and physiological ailments (pestilence, boils) to despoil the land and live-stock and to bring bodily and economic suffering to humans.

In the Covenant Curses

This same perspective is evident in the curses embodied in the law, or the threatened results of breaking covenant with God, Israel's *suzerain* ("feudal lord"; see Lev. 26; Deut. 28). When divine judg-ment occurs, the nonhuman creation will also be affected. God will withhold the rain and send scorching heat, resulting in drought and famine (Lev. 26:19–20; Deut. 11:17; 28:22–23). God will send dis-ease (blight and mildew) against crops and animals (Deut. 28:22). Plant-eating insects, such as locusts and worms, will destroy the crops (Deut. 28:38–39, 42). Wild carnivorous animals will stalk and devour both humans and livestock (Lev. 26:22).

In the curses of judgment, too, nature functions in dual roles. Not only will nature be an object of judgment with the land, but its crops and other vegetation and livestock will be devastated. Nature's wild animals, insects, violent weather, and diseases will also be the agents of God's judgments upon wayward Israel.

In the Prophets

As the prophets applied the curses of the law to their contemporary situations, they also painted pictures of God's judgment against God's own people and against other nations in terms of ecological disaster. The land would become a barren wilderness without crops, humans, or domesticated animals, and would be polluted with dead bodies (see Isa. 24; Jer. 4:23–27; 9:10–14; 12:4; 50:3, 13, 38–40; Ezek. 6:14; 12:19–20; 29:9; 35:3–4, 7–9; Hos. 4:1–3). The prophets were explicit; this would be the result of human disobedience.

As in the Genesis flood narrative, the picture of judgment painted by the prophets is that God would deconstruct his own creation, bringing it back to the situation of Genesis 1:2—formless (*tohu*), empty (*bohu*), and shrouded in darkness. Both Isaiah (Isa. 34:11) and Jeremiah (Jer. 4:23) employed these two Hebrew nouns (*tohu* and *bohu*), famous from Genesis 1:2, to describe the earth's condition subsequent to divine judgment. Only in these three biblical passages do these two terms occur together. These prophets intentionally portrayed the earth after a divine judgment as being in a de-created condition.

The prophets predicted, and some of them experienced, the invasions of foreign armies as God's primary means of bringing about this de-creation through judgment. Jeremiah 12:7–13; Ezekiel 30:10–12; and Habakkuk 2:16–17 are but three examples of many we could list. But the ecological disasters of judgment also occurred through elements of nature, such as drought, wild beasts, locusts, and plague (see Ezek. 14:13, 15, 19; Joel 1:4–12). The prophets also spoke of God's judgments as accompanied by meteorological phenomena in which the sun, moon, and stars will be darkened (see Isa. 13:10; Jer. 4:28; Joel 3:15; Amos 8:9). This, too, is an act of de-creation, reversing and undoing both the separation of light from darkness of creation day one (Gen. 1:3–5) and the creation of the celestial lights of creation day four (Gen. 1:14–19). Thus, the prophets also presented nature, not only as affected by God's judgment directed against human disobedience, but also as God's agent in bringing it about.

In the Apocalypse

Similarly, within the book of Revelation, nature is both drastically affected by and used to administer God's judgment against sinful humans. A primary example is the judgment of the seven bowls of chapter 16. The pouring out of the second and third bowls turned the sea, rivers, and even the springs of water to blood, killing every living thing (vv. 3–4). With the outpouring of the sixth bowl, the Euphrates dried up (v. 12). At the seventh bowl, a severe earthquake leveled the earth (vv. 18–20).

Following our now known familiar pattern, not only was nature impacted, but God also used nature to afflict humans directly. In the sequence of the seven bowls, worshipers of the beast's image were struck with diseases (v. 2); the sun produced scorching and searing heat (vv. 8–9); darkness enveloped the kingdom of the beast (v. 10); and lightning, an earthquake, and hailstones collapsed the cities and killed the people (vv. 18–21).

Within these four different biblical literary genres—historical narrative, law, prophetic literature, and apocalyptic literature—we find a consistent and dominant dual theme. First, specific ecological disasters can be the result of God's judgment. Nature innocently suffers in the judgment evoked by human sinfulness. This highlights the interrelatedness between humans and creation, and between the moral and physical realms. What humans do in the moral realm against God and their fellow human beings affects nature in the physical realm.

But second, nature is not only the unwitting victim of judgment evoked by and directed against humans. Nature also is a divinely employed agent. Natural elements and catastrophes can be the means through which God administers the punishments of judgment—justice upon humans.

Summary and Conclusion

Adam and Eve's sin disrupted the original creational harmony—the interconnectedness between the moral realm in which only humans (of this earth's creatures, at least) function and the natural realm of the rest of this creation. So we humans need to recognize that we are fundamentally the cause of the messed up environmental condition of the world, whether that be due to the first act of sin in the garden or to our own specific sinful actions.

When attempting to explain a particular ecological disaster theologically, however, discernment is required. Any given event may be due to God's executing judgment against humans, judgment that both disrupts and employs nature. Or, it may be caused by the intentional actions of humans—our overt abuse of nature. Or, it may be due to human actions that have unintentionally and unwittingly produced destructive environmental changes. Or, it may be due simply to the fact that natural disasters are just part of the way the world functions in its fallen condition.

Discerning theological causes underlying ecological breakdown helps us to respond correctly and effectively. We may need to repent of sins against God, against one another, and against nature. We probably will need to alter our behavior in appropriate ways. When disasters such as hurricanes and tsunamis devastate our neighbors—whether next door or around the world—we will need to respond wisely, through aid efforts to be sure, but perhaps also by exploring ways to prevent similar future events from inflicting such catastrophic death and destruction. All this, and much more, we should see simply as responding to God's creation mandate to exercise a protective and beneficial stewardship dominion over nature.

Suggestions for Reflection and Action

1. Read Genesis 3, noting in your own words how Adam and Eve's first sin disrupted the various harmonious relationships within the original creation.

2. By yourself or as part of a group exploring these issues, study the passages listed in this chapter and others as you find them. In each, look for the causes given for the reported or threatened disaster. Note how human sin and God's judgments affect not only humans, but also the environment.

3. Consider (and discuss in a small group, if you can) how the four theological perspectives on explaining specific ecological disasters may help us understand and respond to natural catastrophes, such as the Southeast Asia tsunami of 2004, Hurricane Katrina of 2005, major forest fires in the American west of 2008 and 2009, or the Haitian earthquake of 2010.

4. Consider whether any of your own personal actions are in some way abusing or not caring for the creation God has entrusted to us to steward properly. If they are, what changes could you make?

For Further Reading

Anderson, Bernhard W., ed. *Creation in the Old Testament: Issues in Religion and Theology*. Philadelphia: Fortress Press, 1984.

This collection of essays on the theme of creation in the Old Testament concludes with a chapter called "Creation and Ecology." Now almost thirty years old, these are classic essays on the issues discussed in this chapter.

Fretheim, Terence E. *God and World in the Old Testament: A Relational Theology of Creation*. Nashville: Abingdon Press, 2005.

Probably the best contemporary work on the comprehensive theology of creation of the Old Testament. Although more scholarly in style, it provides an excellent balanced perspective on God's intentional creative design of the interrelationships between and among God, humans, and the rest of creation.

GOD'S ONGOING REDEMPTION OF ALL CREATION

Christopher T. Bounds

For the creation was subjected to frustration, not by its own choice, but by the will of the one who subjected it, in hope that the creation itself will be liberated from its bondage to decay and brought into the glorious freedom of the children of God. We know that the whole creation has been groaning as in the pains of childbirth right up to the present time.

—Romans 8:20–22

Introduction

Many Christians today are infected with a form of Gnosticism, a heresy that plagued the early church. Gnostics believed our physical world, including the human body, is destined for destruction in the life to come. Only the spiritual will exist in heaven. Gnostics thought earthly matter is, at best, a divine mistake, and at worst, intrinsically evil. The Gnostic concept of salvation, therefore, focused on liberation from the body and birth into a purely spiritual world.

I see this virus up close in the classroom and in the local church. At the university where

> [The world] will then contain no jarring or destructive principles . . . not thorns, briers, or thistles . . . but every [plant] that can be conducive, in anywise, either to our use or pleasure. For all the earth shall then be a more beautiful Paradise than Adam ever saw.
>
> —John Wesley

I teach, I always ask pre-ministerial students in my basic Christian doctrine course what humanity will be like in "the life everlasting," as the Apostles' Creed calls it. Given the choice between a spiritual existence only in heaven and one with a real physical body, most opt for the former in clear contradiction of New Testament teaching on the resurrection of our bodies.

When I attend funeral services, unless historical liturgies are used, I often hear little or no reference to the departed believer's resurrection at Christ's second coming. Instead, the focus is on immediate entrance into heaven and the experience of final reward; the disembodied, spiritual existence of a person in death is elevated to the state of ultimate salvation. From a biblical and theological perspective, however, there is no complete work of redemption without bodily resurrection. To be fully human in the life everlasting means we must have a physical body.

Gnosticism infiltrates our views of the created order as well. As with our physical bodies, many do not see the created order as a part of the life everlasting. While they affirm God's declaration of creation's goodness in Genesis and may recognize the effect of human sin on it, they fail to see the full implications of God's salvation. Christ came not only to redeem humanity from sin, but all of creation as well. Through Christ's ministry, we will realize the ultimate purposes of God along with the created order. Creation participates with us in God's salvation, perhaps only partially in the present, but in the future, fully.

The purpose of our chapter is threefold: (1) to expose Gnostic tendencies latent in our Christian thinking; (2) to examine the riches of God's salvation for all of creation and its implications for the world today; and (3) to explore God's ultimate purpose for the created order, and its participation in redemption at Christ's second coming. We begin by briefly reviewing God's work in bringing creation into existence, and the effects of human sin on it.

Creation and the Fall

The biblical narrative of the created order begins in the opening chapters of the Old Testament. In contrast to other ancient religious texts on how the world came into being, Genesis teaches that God directly created the physical universe with wisdom and purpose and judged it "very good" (Gen. 1:31). Creation is an expression of divine love. As such, there is nothing inherently evil, sinful, or unspiritual about the physical world.

God made humanity a part of the created order—embodied souls in the divine image (Gen. 2:7) with the express purpose of exercising stewardship over it (Gen. 1:28). The *imago Dei* (Latin "image of God") in humanity had a threefold nature: moral, natural, and political. The moral image mirrored the holy love of God and formed the basis of human character. The natural reflected God's rational and creative capabilities, enabling sound human understanding and judgment. Finally, the political image rightly ordered humanity's web of relationships, making possible perfect connection to God, other human beings, and the rest of creation.

Therefore, the divine likeness gave humans a unique place in the world, equipping them with the tools necessary for their responsible care of it. In the garden, holy love informed our first parents' reasoning, understanding, and will, resulting in responsible treatment of the created world, rightly ordered human relationships, and perfect love and obedience to God. All creation existed in harmonious relationship through the holy leadership of our first parents.

Genesis makes clear that while humanity has an exalted position in creation, we are inseparable from it. To be human is to be physical, to be formed of the dust of the earth (Gen. 2:7). Our humanity and our created existence are part and parcel of each other. Because we are physical beings, what happens to us affects the world and what happens in the world affects us. Ours and the world's past, present, and future all are intimately intertwined.

While humanity began in perfect rapport with creation, this changed in a moment. Our first parents rebelled against God and committed sin, resulting in brokenness and alienation in all human relationships (Gen. 3). This fall reversed the original conditions of human life, devastating the *imago Dei* in humanity. Morally, we became unresponsive to God, self-focused, and enslaved to sin. Naturally, our rational capacities and our exercise of will were impaired dramatically. Politically, our ability to relate to one another in healthy ways and to live in responsible harmony with creation ended.

Because of humanity's unique position in the world, our sin touched every part of the created order. Brokenness, corruption, and death marked our human bodies and entered the created order. The world began to behave in ways not originally intended by God. While, in God's providence, it continues to work, it does so with many problems. Furthermore, our ability to exercise essential stewardship in creation is impaired by the brokenness of our relationship with it (Gen. 3:17–19). Blinded to the essential goodness of creation and limited in judgment and will, we began to exploit it for selfish ends, withholding the care it deserves and needs.

Most Christians recognize the basic points of creation and fall in the biblical narrative, and the need of divine intervention for our salvation. However, the Gnostic virus of which we spoke impairs our vision and our thinking, causing too many of us to miss the necessity of creation's redemption in this drama. Without salvation of the physical world, we cannot be redeemed either. To be human means to be connected to the physical world. The destiny of humanity and the rest of creation are interdependent. God's salvation addresses all creation, not just a part of it.

Redemption of All Creation

Through the work of the eternal Son in his incarnate life, death, resurrection, and exaltation, God has redeemed humanity and creation. Christ did everything necessary for salvation, paving the way for the realization of the ultimate purposes of humanity and the created order, beginning in the here and now, and culminating in the life everlasting. We will discuss three key aspects of Christ's work.

Christ's Incarnation

The Scriptures and the Church both affirm Christ's assumption of human nature—his becoming an embodied soul. We call this understanding the doctrine of the incarnation. Through Jesus Christ, God's nature unites with created existence. The incarnation itself stands as the ultimate declaration of the goodness of creation.

The shock of God's union with physical nature was too much for some in early Christianity. An early Gnostic heresy, Docetism, taught that the Son of God only appeared to come in the flesh. The incarnation was an illusion because it was beneath the dignity of God to bond with any part of the created world, even with human nature. The early church rejected such teaching. Instead, they saw the incarnation as a witness to the testimony of Genesis. Creation's essential goodness made possible the union of divine and human natures in Christ.

Christ's Resurrection

Second, Jesus' resurrected body after death points to our future resurrection. In Christ's resurrection, human nature is transformed, becoming incorruptible, no longer subject to disease, death, evil, and sin. While our physical bodies are subject to decay and death in the present life, we live in hope of bodily resurrection. We follow Jesus in death, and we will follow him in resurrection after death. At Christ's second coming, those who have died will be resurrected;

those still living will be transformed into the likeness of Christ's resurrected body.

However, Christ's bodily resurrection also anticipates the future of all created existence, when God will transform the world and be "all in all" (1 Cor. 15:28). Jesus' glorified body is the sign of creation's future. More specifically, our resurrected bodies must have a physical order in which to live. This is the place Jesus promised he would prepare for us and come again to take us there (John 14:2–3). We are inseparable from the created world. Therefore, just as creation shared in humanity's corruption and fall in the garden, it will participate in the full work of God's redemption, the glorified and incorruptible state of resurrection.

Ongoing Redemption of Creation

Finally, just as our salvation in Christ is both already and not yet, beginning in this life and culminating in the life to come, so also creation presently participates in salvation. Many of us recognize that we experience present salvation through forgiveness of sin and sanctification by the Spirit, but how does the rest of creation experience redemption now?

Through the work of Christ and the Holy Spirit, Christians experience a progressive renewal in the image and likeness of God. While the moral *imago Dei* was destroyed in the fall, this image is partially renewed in all humanity through prevenient grace, significantly restored through the new birth, and can be completely rehabilitated in the present life through entire sanctification.

The moral image enables holy love to inform the knowledge, decisions, and actions of humanity. In relationship to creation, it is through prevenient grace that humanity becomes capable of a semblance of responsible stewardship of the created order. Through the new birth and sanctification, an even healthier stewardship is possible.

Humanity's moral renewal makes possible a more sanctified relationship with creation, but it does not guarantee it. While we may experience the full restoration of the moral image in this life, the natural and political images remain marred, resulting in sins of infirmity from clouded reasoning and mistakes in judgment. In relationship to creation, this can lead unwittingly to abuse, even destruction, of the created order.

However, as we grow in wisdom, knowledge, and understanding, we experience renewal also in the natural and political aspects of the *imago Dei*. We begin to understand more fully the implications of creation's goodness, God's eternal plan for it, and healthier ways to relate to it. Together with our renewal in the moral image, we become equipped to carry out more closely God's original intention to exercise stewardship in creation. Motivated by divine holiness reflected in us, and armed with better discernment and understanding of the created order, we are empowered to make better decisions that result in a better relationship with the rest of creation.

Therefore, as God restores the fully divine image in humanity, the work of reconciliation between humanity and creation deepens. The relationship of curse standing between humanity and the physical world because of sin is being lifted thorough our deepening experience of salvation. Because of the riches of God's grace in salvation and our deepening understanding of God's revelation, we are beginning to realize the potential for sanctification extending to our relationship with the created order in the present life. Godly human care of creation is becoming increasingly possible, through Christ's redemptive work in us.

The Culmination of Creation's Redemption

In Revelation 21:1–4, the apostle John gives a beautiful description of the world's future:

Then I saw a new heaven and a new earth, for the first heaven and the first earth had passed away, and there was no longer any sea. I saw the Holy City, the new Jerusalem, coming down out of heaven from God, prepared as a bride beautifully dressed for her husband. And I heard a loud voice from the throne saying, "Now the dwelling of God is with men, and he will live with them. They will be his people, and God himself will be with them and be their God. He will wipe every tear from their eyes. There will be no more death or mourning or crying or pain, for the old order of things has passed away."

Salvation begins in the present life, but climaxes in the life to come. This final work will not come through any human activity, but only by God's decisive action. Through bodily resurrection at Christ's second coming, final justification at divine judgment, and union with God in the life everlasting, we will realize the definitive purposes of God in our creation and redemption.

The created order will experience its ultimate salvation as well. In the transition between final judgment and life everlasting, the created order will follow humanity in a resurrection-type event. The present order, disfigured by the history of sin, will experience complete transformation, when corruptible existence will take on incorruption. As Peter frames John's statement, there will be "a new [physical] heaven and a new earth" (2 Pet. 3:13). Creation as we know it now will be transformed, and a new age for the world will dawn.

While there has not been real debate in historic Christian teaching about the physical nature of the life everlasting, questions have arisen about the new creation's source. Some theologians have interpreted such words and phrases as "perish" (Ps. 102:26), "be dissolved" (Isa. 34:4), "vanish like smoke" (Isa. 51:6), "pass away" (Matt. 24:35; Rev. 21:1), and "disappear . . . be destroyed . . . melt" (2 Pet. 3:10, 12) as teaching that a new physical order will replace the old entirely. Rather than a transformation of the present world, God will fashion a fresh, brand-new creation, unrelated to this previous one, a new *creatio ex nihilo* (Latin "creation out of nothing").

However, most Christian interpreters, including John Wesley, have disagreed, understanding these biblical passages to point to the present order's transformation, not destruction (Rev. 21:5). "Passing away," "destruction," "consummation by fire," and similar phrases are simply images of purification used to describe the cleansing and renewal of the world, not its annihilation. The present shape of the world is destined for destruction, but not its actual existence.

These theologians have looked to Christ's bodily resurrection as a clue to the type of transformation our physical bodies as well as the rest of the created order will undergo. Reading the New Testament accounts of Jesus' resurrected body, they have noticed: (1) it is physical—body that is seen, touched, embraced, and that can break bread with others; (2) it is connected to his former earthly body, expressing that body's miraculous transformation; (3) it exists in continuity with our present bodies, not appearing to be different from the human bodies around him; and (4) it is not bound by the same physical laws by which our present bodies are bound, having the ability to appear and disappear, ascend and descend.

In addition, Paul taught that our bodily resurrection will be like that of Jesus, whom Paul regarded as the firstfruits (1 Cor. 15:23). Jesus' body was raised incorruptible, no longer subject to death. Just so, our resurrection bodies no longer will be subject to death, nor to disease, sin, and sin's other effects.

Also, we should note that Christ's resurrection was qualitatively different from any other resurrection in history. Jesus raised a widow's son, Jairus' daughter, and Lazarus from the dead during his earthly ministry. Yet all three remained subject to death and decay. Ultimately, all three died again; they experienced only resuscitation, not a real, final resurrection. Taking all together, then, Christ's resurrection is witness that the new creation will be a transformation of the present order.

The essential goodness of creation is born out both by the divine testimony of the Genesis creation reports, and by the incarnation and resurrection of the eternal Son of God. It follows, most Christian theologians have believed, that it would not make sense for God to do away with what is good. God does not destroy what he has created, but redeems and transforms it, converting it from a corruptible into an incorruptible existence.

More specifically, God's purpose in his final act of redemption is to perfect creation. As Paul describes it, creation "will be liberated from its bondage to decay" (Rom. 8:21), will be fully reconciled with God (Col. 1:20), and will be gathered fully unto Christ (Eph. 1:10), enjoying eternal and perfect communion with God. As noted earlier, John described this as the day when "the dwelling of God is with [humans]" (Rev. 21:3).

The new creation will exist in perfect harmony with resurrected humanity. The book of Hebrews teaches that presently the physical world is not fully subject to us (Heb. 2:8). However, it also points to a time in the future when it will be, fulfilling the promise of Jesus that the "meek . . . will inherit the earth" (Matt. 5:5). In the life everlasting, the moral, natural, and political images of God in us, fully and incorruptibly restored, will enable us to have a relationship with the created order motivated by a holy love, reflective of God's character, and informed by wisdom, knowledge, and understanding, leading to perfect harmony in creation. We will fully realize the stewardship dominion God intended in our creation (Gen. 1:27).

In turn, creation will respond to humanity's stewardship. The reconciliation between the physical world and humanity will be complete, manifested in the perfect ordering of creation. The world will work as intended by God without the possibility of flaw. John pictures this as a day when "the sun will not beat upon" us, nor will there be "any scorching heat" (Rev. 7:16). In his Sermon 64, "The New Creation," John Wesley pictured this harmonious state of creation in the following description:

> The horrid state of things which at present obtains, will soon be at an end. On the new earth, no creature will kill, or hurt, or give pain to any other. The scorpion will have no poisonous sting; the adder, no venomous teeth. The lion will have no claws to tear the lamb; no teeth to grind his flesh and bones. Nay, no creature, no beast, bird, or fish, will have any inclination to hurt any other; for cruelty will be far away, and savageness and fierceness be forgotten. So that violence shall be heard no more, neither wasting or destruction seen on the face of the earth. "The wolf shall dwell with the lamb," . . . "and the leopard shall lie down with the kid: They shall not hurt nor destroy," from the rising up of the sun, to the going down of the same.[1]

In final redemption, God's holy love will define all creation and no fall of humanity or creation will be possible. Creation's final form will far surpass the perfection it enjoyed in the beginning, and humanity will be like God fully, in impeccable character. All creation—including humanity, its crown and steward—will be in consummate union with God.

Conclusion

Contrary to Gnostic tendencies in contemporary Christianity, creation has always had a central role in God's eternal purposes and redemptive work. We matter to God and so does the rest of the created order. We are inseparable, and together, we share in God's ultimate will. The created order does not pass away in the "life everlasting." Rather, it is brought to its ultimate perfection and will share in God's glory with us.

Suggestions for Reflection and Action

1. Identify and reflect on a specific moment in your life when the idea of heaven or the life everlasting was presented as only a spiritual existence, either explicitly or implicitly. Did it seem like sound theology then? Does it now?

2. How does (or could) recognizing the significance of creation to God, and to God's eternal purposes in Christ, change how you think about the created order? Ponder this question long enough to arrive at several specific answers. Consider discussing it in a formal or informal small group.

3. Because of our new birth in Christ and sanctification through the Holy Spirit, we experience reconciliation with God. This moves us to work toward reconciliation with the rest of humanity and with the created order. What concrete steps can you take, individually and with others, to work toward reconciliation with the created order? Be specific; work toward putting your answers into action.

4. Write down one insight, question, or action step you are taking away from this chapter. What will you do with it?

For Further Reading

Grider, J. Kenneth. *A Wesleyan-Holiness Theology*. Kansas City, Mo.: Beacon Hill Press, 1994.

In his last chapter, Grider provides a helpful, easy-to-read overview of the Wesleyan/Holiness tradition on eschatology, including the place of creation in God's redemptive plan.

Oden, Thomas C. *Classic Christianity: A Systematic Theology*. New York: HarperCollins Publishers, 2009.

In his discussion of eschatology, Oden attempts to state a Christian consensus on creation's experience of redemption. Through Oden's treatment, the reader will see that John Wesley's teaching on the subject and the Wesleyan understanding of it are shared by most of Christianity.

Wesley, John. Sermon LXIV, "The New Creation," and Sermon CXXXVII, "On the Resurrection of the Dead," in *The Works of John Wesley*, Kansas City, Mo.: Beacon Hill Press, 1979.

These two sermons highlight the major points of Wesley's belief in creation's full redemption in the future kingdom of God.

Wright, N. T. *Surprised by Hope: Rethinking Heaven, the Resurrection, and the Mission of the Church*. New York: HarperOne, 2008.

Wright addresses the Gnostic tendencies affecting contemporary understandings of creation's present and future participation in God's redemptive plan. He grounds his discussion both in biblical exegesis and in the church's historical understanding.

Note

1. John Wesley, "The New Creation," *The Works of John Wesley*, vol. II (Kansas City, Mo.: Beacon Hill Press, 1979), 295.

GOD'S CONSTANT CARE OF THE UNIVERSE

Kenneth F. Gavel

For all things in heaven and on earth were created by [Christ]—all things . . . through him and for him . . . For God was pleased to have all his fullness dwell in the Son and through him to reconcile all things to himself by making peace through the blood of his cross—through him, whether things on earth or things in heaven.

—Colossians 1:16, 19–20 NET

Introduction

Caring for creation has become a hot-button issue for many evangelical Christians, although not all agree concerning the extent of our current responsibility for the environment. The spectrum ranges from those who argue strongly for greater Christian involvement to those content to evangelize and leave the fate of the earth in God's hands. (In keeping with the scope of this book, we will not always distinguish here between Christian care for people and Christian care for the rest of creation, but will include both under the rubric of "creation care.")

> All things by immortal power,
> Near or far,
> Hiddenly
> To each other linkèd are,
> That thou canst not stir a flower
> Without troubling of a star.
>
> —Francis Thompson

From a biblical perspective, we understand that we have a mandate to care for our world (see chapter 1), that we have ourselves to blame for the current crisis (see chapter 2), and that, ultimately, God will redeem and restore all (see chapter 3). The question remains, then, does God expect us to play a redemptive role now? To explore our place in this drama, we will revisit our God-appointed place in the world—the issue of the Christian's present relationship to God and God's world.

Whose World Is It?

Christians understand all of space, time, and matter as the unique creation of the one triune God. Furthermore, God is in no way dependent upon creation, either for his being or for his actions. We can draw three implications from these facts.

First, because God is the sole, independent creator, it is possible to view the world as just the world, a truth sometimes expressed in the phrase, *creatio ex nihilo*—creation out of nothing. The relative independence of this world is ensured by God's faithful sustenance of its properties and functions through the laws of nature. (I use *law* here as shorthand for those labels given to observed regularities of the universe. I do not mean to suggest a set of inflexible rules that God must break to act directly in space-time reality.) These laws do not require any human contribution. Neither do they restrict God from acting more directly (by miracle) in the world. They do ensure, however, that humans have enough independence to act freely and responsibly.

God's provision of basic, reliable laws for a rightly functioning creation ensures a high level of stability for our lives. However, these laws also allow for a high degree of contingency and possibility—hence, for a vast variety in life. The result is both high privilege and staggering responsibility.

Second, because together they are committed to it, all three persons of the triune godhead—God the Father, God the Son, and God the Holy Spirit—continue to be intimately involved in this world.

Thus, we must not restrict ourselves to seeing the world as just the world. We must see it also through the lens of redemption. As noted in the previous chapter, the story of creation is not yet in its final form.

Third, because God is sole creator, humanity is not the center of the story, as some teach. This means that life on this planet, even human life, is not the most important reality; God is. If our focus is primarily on environmental and human health, then we are essentially eco-centric, cosmo-centric, and anthropocentric. As faithful Christians, our focus instead should be on God, and on our participation in God's agenda. The whole of creation belongs to God (Ps. 89:11). This is why our theology of creation is so important—it helps us keep the right focus. This is why the first section of this book is devoted to biblical theology. This is also the reason some evangelicals prefer to talk about creation care rather than environmentalism.

Precisely because God is at the center, we concern ourselves with God's intentions and purposes for this world. In so doing, we discover that God's intentions require the involvement of humans.

God Re-Stakes His Claim

Nowhere is the divine ownership, together with God's intentions for creation, expressed more powerfully or defined more clearly than in the incarnation of the second person of the trinity. It was no accident that the Son took human form. It was through the sin of the first Adam that death entered the created order (Rom. 5:12). However, it is through the second Adam (the gift given to us through God's grace) that many shall be made alive (Rom. 5:15–17; 1 Cor. 15:22).

Grace within Incarnation

God's grace has two basic meanings: grace as attitude and grace as power. Grace as attitude refers to God's favorable disposition toward us, as a result of Christ's atoning work, by which God is able to forgive us. It has a forensic connotation and result—a legal declaration of pardon.

Grace as power (the second sense) refers to God's transforming work within, to the making-holy work God does in one's spirit and character. By this grace we become new creatures in Christ (2 Cor. 5:17). By this grace we bear the fruit of the Spirit (Gal. 5:22–23), which is the character of Christ. We do serious injustice to the work of Christ if we reduce it to only a matter of God forgiving our rebellion. Even more importantly, God's grace is manifested in our lives as the transforming power of the Holy Spirit. For this reason, we rightly link holiness to wholeness.

A biblical view of the incarnation understands Jesus' mission as the restoration of all creation to its designed purpose: to redeem and transform sinful souls, to re-create physical bodies, to restore social structures to a state of harmony, and to re-create the physical world. The one through whom all created things were made (John 1:1–2) also is the one through whom all things will be re-made (Col. 1:15–20). For this reason, we may speak not only of saving grace in a spiritual sense, but also of graced creation.

The incarnation lets God remain God, and enables humans to regain our identity and purpose, a part of which is to care for creation. The incarnation brings about God's original intent: to create a holy people filled with his Spirit with whom he would dwell in intimate fellowship as they perform the tasks assigned to them. Hence, the incarnation is about much more than merely getting our souls to heaven. It is about reclaiming a creation corrupted through human sin. It affirms the original goodness of the physical creation order. It shows that God refuses to abandon the project. It shows that just as God makes us fellow workers in spreading the gospel (2 Cor. 5:18—6:1), he also makes us partners in preserving the environment in which all earthly life thrives.

Facilitating wholeness and reflecting God's glory to the rest of creation, Christians become ministers of rehabilitation and re-habitation. Jesus was serious in labeling his followers "salt of the earth" and "light of the world" (Matt. 5:13–14). As salt, those who participate in the character of Christ help awaken the senses of those around them to the joy and true meaning of life. As light, they demonstrate through their practical works (Matt. 5:16) love for fellow humans and reverence for God's good creation. Christians, above all others, ought to demonstrate the mindset and attitude of restoring harmony with the creator and his purposes for creation, for in Christians, God's kingdom rule has indeed begun. As Manfred Marquardt puts it, Christians represent a counter-reality in a wicked and broken world, "ready to be peacemakers and comforters, to bring justice and solidarity, to live out the love they receive daily from their Creator and Sustainer."[1]

Participation within Providence

A fundamental rule of creation and of God's ongoing care of it is that God enables the creation to realize its potential. To allow all aspects of creation to have the real dignity of being distinct from God and to allow them to make a real contribution to the creation story, God has chosen not to do by special miracle what God can accomplish through ordinary means. The glory God desires is that manifested by God's shining through creation. The love within the trinity is so abundant and perfectly fulfilling, it is free to create an "other" and allow it to have a real difference from the Creator. God receives the greatest glory by invisibly enabling each part of creation to fulfill its unique, God-given purposes.

This principle of providence, which allowed Adam and Eve to make a real contribution to the harmony of the original creation, is threatened in a fallen creation. This fallen creation raises the specter of evil and evokes questions about the wisdom and power of God's care of the world. That the problem of evil arises at all as a moral and philosophical category is due precisely to the fact that God has not changed his mind about this principle of creation and providence. Therefore, while we prefer quick fixes, God prefers responsible participation. While we prefer miracles, God prefers costly process. We prefer the fast-food miracle of the loaves and fishes (John 6:1–14); God prefers that we learn to grow wheat and make bread, make nets and catch fish. We prefer to wait for the miraculous new heaven and new earth in the future; God requires that we care for our polluted rivers and oceans now. We are often content to wait for justice to be meted out at the great white throne of judgment; the love of God infused into our hearts by the Holy Spirit requires that we rescue and restore abused creation, both human and non-human, now.

Of course God can and does work miracles, but God did not create this world as a magical kingdom. Just as God does not miraculously spoon-feed us our porridge, just as God does not plant and harvest our corn for us, so God does not miraculously pick up our garbage, preserve the fish from our careless disposal of chemical waste, or protect livestock from pesticide poisoning. Paul expressed this providence-principle in the admonition, "Anyone unwilling to work should not eat" (2 Thess. 3:10 NRSV).

Responsibility within Restoration

The point is, God has delegated the great dignity of being human to us. With that dignity comes responsibility, a dignity and responsibility that God in his gracious love refuses to withdraw. To absolve us of responsibility for our neighbor's spiritual and physical well-being would be to abort the creation project, to abort the human project.

We need to see ourselves, therefore, as an intrinsic part of the world—our welfare and that of the natural world around us are closely intertwined. This fact has both a biological and biblical basis: We share in the common dust of the earth; the broken, non-human creation waits for its own final redemption when the children of God will be manifested (Rom. 8:19–21).

Not only does God grace us with physical life and new spiritual life, but he also graces the rest of creation through us. God is indeed with us, but it is to empower us, not to stand in as a substitute.

God provides the basic underpinnings of life—wild animals are able to adapt to their environment; birds find the food they need (Matt. 6:26). God eventually will re-create the natural environment to provide for those needs in such a way that there will be no want, suffering, sorrow, or death. In the meantime, we are mandated to participate, as far as possible, in reclaiming and fostering life in all its dimensions on this planet.

We may summarize by noting that humans are uniquely privileged and tasked by God with a creation mandate, a biological mandate, a redemption mandate, and a moral-ethical-love mandate to care for our environment in all its complexity.

God privileges us to be coworkers in the ministry of reconciliation (1 Cor. 3:9; 2 Cor. 5:16–20). If God considers us capable of partnering in that high calling, it should not be surprising that God does not treat us as lifeless machinery. God expects us, as temples of God's Spirit (1 Cor. 3:16–17), to do our own cleaning and routine maintenance. This includes, for example, taking care of our health by regularly brushing our teeth and cleaning the mildew out of our showers. God also expects us to take care of our larger earth-home by keeping it clean and free from conditions and substances that lead to disease and death.

To this end, God has endowed us with intelligence through his creative design; empowered us with love through his re-creating Spirit; and included us in a community of communities—with God, with humanity, and with all nature, through all of whom he diffuses *shalom*. Our participation in creation care is a practical extension of our responsibility to bring the good news of redemption and rescue to as much of creation as possible. Creation care is compassion in action, our participation in extending God's intended harmony to all corners of his good creation.

All this being true, there should be a direct connection between a Spirit-filled Christian and genuine care for neighbor and environment. Wesleyan theology is founded on a doctrine of holiness that by definition strives for wholeness in all dimensions. Precisely because God has intervened in the incarnation, we can be in right relationship with him and, hence, in right relationship with our environment—not only with our fellow humans, but also with all of God's good creation. Having received the firstfruits of God's redemptive and re-creative grace, we also have a mandate to set the standards in creation care.

We may be tempted to jump up and rush off in all directions at once. But a further word is in order, a caution or a precondition that will keep us from a foolish waste of energy. We will act with wisdom only if we remember that the practical how-to does not work without the spiritual enabling. In biblical and spiritual terms, one cannot have the character of Christ and remain at ease with threats to God's *shalom* in the world. One cannot love one's neighbor without participating in God's kind of love.

Thus, the unconditional beginning point of caring for God's creation is to obey his call to separate ourselves absolutely to him. Only then can we be free of the self-centeredness that destroys both relationships and the environment. And only then can we be filled with the Holy Spirit and energized by the "infectious love of God."[2] And only then can we participate in God's agenda for the world and fulfill our areas of responsibility.

Historical Mentors in Creation Care

A popular, widespread assumption is that Western Christianity has always devalued the environment or at least has valued it only insofar as it serves the interests of humans. This assumption is, at best, an overstatement. Over the centuries, many Christians have raised their voices, not only in support of human welfare, but also for the physical environment, animals, and creation care in general. Our own historical tonedeafness sometimes prevents us from hearing them.

Prior to Wesley

It is true that Christians have not always practiced proper stewardship of the world around us. Origen and Thomas Aquinas, for example, tended to focus on God as "the Creator of all things, but not the redeemer of all things."[3] However, Irenaeus, Augustine, and St. Francis of Assisi understood the intrinsic connection between love for our creator and provider and love of others and the rest of creation. Moreover, while the reformers Luther and Calvin tended to focus on human salvation, they did see nature as "the theater of God's glory."[4]

In an historical overview, Paul Santmire discusses two "fundamental ways of thinking in Western theology" concerning nature: (1) a "spiritual motif," with a primary focus on humans as spiritual beings called to rise above the mundane to find ultimate communion with God—such as, an otherworldly focus; and (2) an "ecological motif," which sees humans existing within "a system of interrelationships between God, humanity, and nature."[5]

Historically, the most famous representative of the ecological motif was the thirteenth-century St. Francis of Assisi, renowned for his love of animals and the physical universe. Laurel Kearns cites these attributes of Assisi as the reason "Pope John Paul II named him the patron saint of ecologists."[6] Many evangelicals are also glad to claim Assisi's life and theology as precedent for their championing of this ecological motif in their current calls for creation care.

John Wesley

John Wesley (1703–1791) sought a return to a biblical Christianity, which he defined as the experience of a heart transformed and empowered by God's love. For Wesley, the powers of the world to come have broken into this world already through Christ and his kingdom. This kingdom of God has been set up in the heart and life of every true follower of Christ.

In the matter of absolute consecration and being filled with the Holy Spirit, Wesley set an example of a holistic love embracing both the spirit and the body, both spiritual needs and social, political, and economic needs for one's neighbor and for oneself. Famously, this biblical focus motivated Wesley to develop a myriad of practical ministries to, for, and with the poor.

These ministries included many well-documented forms of direct assistance. Not so well-known are Wesley's and early Methodism's provision of short-term loans (often today called micro-loans) and of work for destitute women. Moreover, Wesley was an early advocate for the abolition of human trafficking, in his day simply called slavery.

Wesley's concern for the whole person gave him a deep interest in the medical arts, healing, and in the natural sciences. He had a special interest in astronomy, since the heavens reflect God's creative power, wisdom, and goodness.

Behind this focus on the mundane matters of the physical universe lay Wesley's understanding of grace. God's grace given to us in Christ and through the Spirit is empowering and liberating. Therefore, although God must always take the initiative, it is precisely because of this initiative that all humans have a responsibility to use the grace we have to do all the good we can. As we do so, we receive more grace. Thus, Christians are not just passive recipients of God's pardon; we are also participants in God's character and works. With empowerment comes ability and, therefore, responsibility toward all aspects of God's creation.

Wesley's interest in both revealed and natural theology led him to think of all aspects of God's creation as being interrelated, with each level of creation having its own unique purpose and right to exist, quite aside from any utility it may or may not have for humans.[7] Thus, we see in Wesley the incorporation of godly love of others and care for God's world into the Christian love of God.

Conclusion

The triune God created the universe. God will not let it fall apart. Paul declared that in Christ "all things hold together" (Col. 1:17). Yet, God entrusts humans with stewardship responsibility for this planet.

With Wesley and many others, we understand that because humans are both spirit and body, we cannot restrict our interest only to the spiritual. To love God is to love our neighbor—spirit and body (see 1 John 3:10–18). To love our neighbor is to care for our common environment. Anything we do to make the plant and animal world better is loving our neighbor, since the earth's complex ecological systems make up the environment in which life flourishes. Moreover, since the environment—the earth—is creation, to slight it is to slight the creator. Matter matters.

The very name *Christian* ought to make it a foregone conclusion that Christians will take an active role in caring for God's creation. The love and compassion of Christ, breathed into us by the Holy Spirit, can do no less. God has honored us with the incalculable privilege of working together with him in bringing his *shalom* to our broken world.

Suggestions for Reflection and Action

1. Reflecting on these environmental issues, assess your present response. Is it avoidance? Engagement? Despair? In light of your developing biblical eco-theology, what do you think a Christian mindset may be, individually and together?

2. Craft a plan for developing a biblical theology of creation stewardship and implement it. This may be easier with a partner or with a small group of others who are reading (or have read) this book.

3. Based on what you are learning, what practical changes might you adopt to live as the graced participant discussed in this chapter? What two or three behaviors could you change relatively easily within the next week or two?

For Further Reading

Bouma-Prediger, Steven. *For the Beauty of the Earth: A Christian Vision for Creation Care.* Grand Rapids, Mich.: Baker Academic, 2001.

This award-winning book provides a theologically sound and ecologically informed call to Christians to own their stewardship responsibilities.

Dewitt, Calvin B., ed. *The Environment and the Christian: What Does the New Testament Say About the Environment?* Grand Rapids, Mich.: Baker Publishers, 1991.

As the title suggests, this book surveys New Testament teaching and attitudes toward the environment.

Kearns, Laurel. "The Context of Eco-Theology." In *The Blackwell Companion to Modern Theology*, ed. Gareth Jones. Malden, Mass.: Wiley-Blackwell, 2004, 2007.

Kearns gives a brief overview of the history of Christian attitudes to nature, including the breadth and type of current interest. She provides a good introduction to key figures, and suggests her own typology of Christian responses to environmental concerns.

Santmire, H. Paul. *The Travail of Nature: The Ambiguous Ecological Promise of Christian Theology.* Philadelphia: Fortress Press, 1985. Santmire's historical overview of Christian attitudes toward the natural world has become a standard text in the field. Though now a bit dated, it still provides a respected assessment of the subject. Two Web sites also are worthwhile. The Evangelical Environmental Network and *Creation Care* magazine, providing helpful resources for study groups, are at http://www.creationcare.org. An address providing links to creation care Web sites sponsored by various denominations is http://www.earthcareonline.org/creation_care_websites.pdf.

Notes
1. Manfred Marquardt, "The Kingdom of God and the Global Society" in *Wesleyan Perspectives on the New Creation*, ed. M. Douglas Meeks (Nashville: Kingswood Books, 2004), 172.

2. Ibid., 163.

3. H. Paul Santmire, *The Travail of Nature: The Ambiguous Ecological Promise of Christian Theology* (Philadelphia: Fortress Press, 1985), 179.

4. Ibid., 179–180.

5. Ibid., 9.

6. Laurel Kearns, "The Context of Eco-Theology," in *The Blackwell Companion to Modern Theology*, ed. Gareth Jones (Malden, Mass.: Wiley-Blackwell, 2004, 2007), 469. Kearn's chapter is a helpful overview of the history of Christian attitudes to nature, including the breadth and types of current interest. Kearns's own label for this ecological motif category of evangelical Christian environmental activism is Christian stewardship. Kearns also includes in her discussion two other categories, eco-justice and creation spirituality; see pages 477–479.

7. Randy L. Maddox, "John Wesley's Precedent for Theological Engagement with the Natural Sciences," *Wesleyan Theological Journal* 44, no. 1 (Spring 2009): 52–53.

PART 2

CARE FOR HUMANITY

THE ETHICAL CHALLENGES OF GENETIC ENGINEERING

Burton Webb and Stephen J. Lennox

So God created man in his own image, in the image of God he created him; male and female he created them. God blessed them and said to them, "Be fruitful and increase in number; fill the earth and subdue it. Rule over the fish of the sea and the birds of the air and over every living creature that moves on the ground."

—Genesis 1:27–28

The world around us is the mighty volume wherein God hath declared himself.

—John Wesley

The Science of Potential

When I (Webb) first started working with cells, I was in my junior year at Olivet Nazarene University. I was fortunate enough to work in the research laboratory of Dr. Richard Colling, and the first day in lab he said, "Come here, and take a look at this!" I walked over to what looked like an upside-down microscope and peered through the eyepiece. There, floating in suspension were very small, round, translucent balls. It was a moment suspended in time. I didn't know it then, but I would become enthralled with these little balls of wonder

> For me, as a believer, the uncovering of the human genome sequence held additional significance. This book [the genome] was written in the DNA language by which God spoke life into being.
>
> —Francis Collins

and invest the rest of my life pursuing a better understanding of how they worked and what they could do.

The cells we looked at that day were cancer cells taken nearly fifty years ago from a woman named Henrietta Lacks. She passed away many years ago, but her cells have been preserved and live on in research laboratories all over the world. I joke with my students now that there probably are more of Henrietta's cells now than when she was physically alive. Suffice it to say, I was hooked. The amazing potential of cells has inspired scientists for decades. Millions of scientific papers have been written about the inner workings of cells, and yet we barely have begun to understand how they work.

We all start out as a cell—two cells, really—a sperm and an egg. Each carries approximately one half the information needed to make a human being. But these cells cannot live long on their own; our bodies make and throw them away on a regular basis. Sperm cells are made by the millions each day. Egg cells are made and stored in the ovaries only to die there or to be released and die on the way out of the body. Separately, they are nothing; they have no real potential. But together . . . !

From the time an ovary releases an egg, it has twenty-four hours to meet up with a sperm cell or die. Egg cells are fragile; they cannot be frozen or stored. Sperm cells, on the other hand, are durable; they can live an amazing seventy-two hours after release. Sperm are easily frozen for use years, even decades, later. Still, neither egg nor sperm can divide. Each is the end product of a long and complicated process called differentiation.

Cell Potential

Under natural circumstances, somewhere in the upper third of the fallopian tube, sperm and egg get together. Fertilization causes several profound and nearly instantaneous changes in the egg. First, the outer layer of the egg becomes impenetrable to other sperm. One sperm, one egg—that's the rule. Then, a series of biochemical events takes place, resulting in the fusion of the genetic material of the sperm with that of the egg. In humans, the result of this genetic fusion is a new cell called an embryo.

The newly formed embryo may look quiet and still, but its quiescence is deceiving. Within a few minutes, it undergoes its first of many cell divisions. Early in this phase, it appears that all the cells have the ability to become any part of the human or a completely separate human. We believe that a separation of the cells into two balls very early in development is what results in identical twins. Within a few hundred cell divisions, the embryo resembles a ball of cells that is hollow in the middle. This ball of cells, called a blastula, actually is not homogeneous at all. Significant changes are taking place inside the cells that make up the blastula, causing them to commit to becoming one kind of cell or another. The process of differentiation has begun; it will not cease until the organism dies.

During differentiation, cells interact with one another, determining (by a process we do not fully understand) that, for example, one cell will become part of the nervous system, while the cell next to it will become part of the skin. The pre-nerve cell, beginning to regulate its own genetic sequences, will turn all its internal efforts into becoming a nerve cell. Some genes will be activated, others deactivated, all with the goal of making the structures and proteins necessary for a fully functioning nerve cell.

If all we made were nerve and skin cells, the process would be amazing. But humans are made of dozens of different cell types, each with unique jobs in the body. Take a close look at your skin. This complex organ is made of keratinocytes, granulocytes, neurons, endothelium, and epithelial cells—just to name a few! Each of these cells carries the same genetic material, but certain bits have been activated, others deactivated, resulting in these uniquely functional cells. Differentiation is cellular potential!

Types of Differentiation

Most cells in the adult body are terminally differentiated. This means they have committed to becoming one type of cell and gone so far down that pathway they cannot become anything else. Terminal differentiation results in the body's functional structures, for example, heart, lungs, and brain. Partially differentiated cells, having gone only partway down the path of development, can become a limited number of things, not just one organ only. Few, if any, cells in the adult body are completely undifferentiated, but two circumstances may cause undifferentiated cells to arise—cancer and a few laboratory experiments.

The potential for what the cell will become lies in the nucleus. There, under the tight control of a group of genes known as homeobox (HOX) genes, DNA is altered chemically. These alterations allow for groups of genes to be expressed or inhibited, thereby changing the structure and function of the cell. However, we do not yet understand much about these processes, and it is just here that we bump up against the theological. How can we learn more about the process of differentiation unless we study it? Could we find treatments for diseases like Parkinson's or osteoporosis if we understood more? What would God have us do with the untapped potential of the cell? What is ethical? What is reasonable?

Genetic Engineering

Another kind of potential for cells consists in what we already can do. We now regularly insert genetic material into cells, forcing them to make things useful to us. Genetic engineering emerged in the 1980s as a powerful tool for understanding the inner workings of the cell. By the 1990s, we had begun to insert genes into bacteria, plants, and animals, causing them to make proteins they had never made before. While this may sound strange to some, many diseases have been treated successfully using these techniques—major diseases, including diabetes, cancer, and many others.

Few people worry about causing bacteria or yeast cells to produce insulin for diabetics, but some worry about taking these same techniques out of the lab and into the world. Plants have been bred for resistance to pests for millennia, but now we can insert genes into our plants that make them truly pest-resistant. We can supply vitamins essential to children's healthy development in plants that formerly didn't carry them. We can increase the quality and quantity of proteins in our foods. We do all this by manipulating the way plant cells grow and develop, that is, by manipulating the expression of the genes in their nuclei.

Of course, genetic material has been transferred between animals and plants for millennia. Fertilization and cross-pollination in plants provide for significant genetic diversity. Hundreds of thousands of virus species are capable of capturing bits of DNA and transferring them between unrelated species. In recent years, our ability to harvest these techniques from nature has proven incredibly effective in teaching us how cells function at the molecular level. The real potential here lies in which questions we will choose to ask, and then in what approaches we will take to answering them.

For example, millions of people around the world suffer from iron-deficiency anemia. They do not get enough iron from their foods, primarily because iron-rich foods are expensive and not available to them. How do we get iron into their diets? One answer may be iron-rich rice. Recently, three Swiss scientists reported that by inserting two genes into a rice plant, they increased the iron in the rice almost one-hundred-fold. Health and environmental concerns about such research always will surface, and they must be taken seriously. However, we also must take seriously our calling to feed the hungry. Good stewardship demands a response. But where are the Christians? So few go into basic science that, in only a few years' time, we may not have much of a voice in these discussions.

Frozen Embryos

Now for the real sticky wicket: what to do with the millions of frozen embryos lying around in fertility clinics across the world? My answer? It's really complicated! Here are the two sides as we see them:

1. Frozen embryos are baby people and should be afforded all the rights of personhood. Many, including most Christians, are against abortion generally and do not believe taking even an embryonic life is a good thing.

2. However, what are we doing for these little lives? If we leave them on ice (literally, in liquid nitrogen), eventually they become non-viable; they will die. If we thaw them, they must be implanted at just the right point in a woman's cycle, or they will die. Most of those we implant will not attach; these, too, will die. How is it different, then, to thaw and disrupt them for research purposes if they will die anyway?

The Ethical Questions

In some ways, we are like explorers stepping out of our ship onto uncharted land. We are excited at the good waiting to be discovered here, but we also feel terror for the dangers lying just beyond the shoreline. Now, straddling exhilaration and fear, we must remind ourselves of two fundamental truths: first, humans are charged to act as God's representatives on earth, and second, humans are not God.

Humans as Stewards of Creation

As soon as God created human beings, God gave us a job to be stewards of creation (Gen. 2:15). As stewards, we are to take care of God's world as God's representatives. Very soon, however, we forgot we were stewards and thought we were owners. We did with the earth as we wished, exhausting many of its resources. The resultant shambles in many places testifies to our failed stewardship.

Dominion or mastery, however, does not mean abuse and destruction. When you master a language, for example, you become familiar with every aspect of it, fluent in it, able to utilize your knowledge of it to full advantage; you fully understand that language. The one who masters a language becomes the one least likely to abuse it.

Similarly, God placed us on this earth to understand it fully. This is why God created the universe to be orderly, and why he made it operate by observable laws that humans can study and comprehend. For this reason (among others) God gave us the gifts of mind and curiosity. God did not just command us to understand; God put us in an understandable place as understanding people.

What does this have to do with the new, largely undiscovered world of genetics? We need to remember that we cannot climb back onto the ship and sail to familiar waters. We have been put on the earth to understand it; this includes the building-blocks known as genes. One could say we were put on this earth to play God.

I'm serious when I say that, for we do play God when we carry on God's work. Moreover, the commission to subdue the earth came before the first sin. Sin's arrival made the earth unruly, but God had a plan to undo the effects of sin. His boldest stroke was the cross, but God was active both before and after that moment. God promises an end to the curse; the new heavens and the new earth will be sin-free (see Rom. 8:19–23; Rev. 22:3).

When we take up our assigned role as stewards, we also take up God's work to reverse the effects of sin, effects like disease, ignorance, animosity, injustice, sickness, loneliness, deprivation, suffering, heartache, and the like. When humans eradicate diseases, accomplish medical miracles, lower infant mortality rates, and raise the literacy of a population, we push back the effects of sin's curse on the earth. Such work does not replace evangelism, but it does represent what it means to be stewards of God's creation. We cannot finish the work of pushing back sin's effects—that will happen only at Christ's return—but he wants us to continue this work until that day. That is why he put us here, to continue his work, to play God.

Playing is the right verb, too. Watch a child at play and what do you see? You see a child imitating an adult. You also see children being children, doing what they are meant to do, for childhood is a time for play. Our calling is to be stewards of God's creation. When we do what we were meant to do, when we imitate God by doing God's work in the world, we, too, truly can be said to be at play.

Some objections go beyond the choice of words. You may protest that since we have been corrupted by sin, we no longer can be trusted with this stewardship responsibility. True, fallen humanity is dangerous, but nowhere in the Bible do we read that sin removed the image of God. Sin has turned our motive from God's glory to our own, but our mandate and gifting are not revoked.

But, you may ask, isn't it risky to take on the task of subduing the earth? Yes, it is risky to step off the ship onto the shore. Stewardship always requires courage, as when your parents trusted you to drive the family car. Safeguards are essential; that is why your parents didn't give you the car keys the first time you asked. God also has built in safeguards to protect against self-seeking. God gives to all humans a sense of right and wrong and an inclination to value the right. God gives us a conscience and places us within families and governments.

These issues are complex and always changing; genetic modification is one important example. Most agree it is wrong to use genetics to produce designer babies. But what differentiates gene enhancement from gene therapy, and how can we maintain this distinction? How can we insure that benefits accrue equitably? One track for answering such questions is the work of groups within the governmental and scientific communities, groups such as the President's Council on Bioethics, the National Academy of Sciences, and the World Health Organization. The difficult questions and the great opportunities mean some within the Christian community should take up the responsibility to subdue the earth by becoming scientists and ethicists, men and women able to speak with intelligence and a Christian voice on these issues to help provide guidelines and safeguards against the risks.

Though the scientific community has just arrived on shore, relatively speaking, already they have proven the risks to be worth it. The gains in easing human suffering have been remarkable. I (Lennox) can speak firsthand. Since beginning on a daily injection of medicine made possible by genetic research, my daughter's quality of life has improved dramatically. Instead of constant pain and debilitation, she now is able to lead a normal life.

Humans Are not God

While made in God's image, humans are not God. To say we are not God means we are under God's authority, not our own. We accept as valuable what God considers valuable, and act according to God's rules.

The Sanctity of Life. As one of many implications, this means we honor the sanctity of human life. No one can say with certainty when human life begins but, given its sacredness, it behooves us to err on the side of caution and assume that life begins at conception. This would mean, for instance, we may support prenatal testing as a way to identify and begin to resolve genetic problems in utero, but do not support the abortion of handicapped little ones.

We recognize the tremendous potential for good that can arise from the use of stem cells, especially embryonic stem cells. But to destroy a uniquely conceived human embryo in order to obtain stem cells is to destroy a sacred life. This is true whether the embryo is aborted, prepared especially for the harvest of stem cells, or left over from fertility treatments.

The question of these left-over frozen embryos is vexing. Remember, they are doomed to die unless implanted. If they are not implanted, some ask whether it is not more respectful to the embryos to use them in research than to let them die. They consider it immoral not to use these embryos, since they offer such great potential for helping others.

While the question is frustrating, we believe the right decision always is on behalf of life. Even if some good can result from destroying an embryo, that potential good never outweighs the good of preserving life. Leaving embryos to die in their frozen state is bad enough, but actively ending their life is worse. Some are choosing snowflake adoption, rescuing these little ones from death either to raise them or put them up for adoption after birth. Even better would be reducing the number of left-over embryos. When Christian couples seeking fertility treatments become aware of the serious ethical implications involved, perhaps they will explore other options.

A corollary to our commitment to the sanctity of life is a commitment to the weakest members of society, those for whom life is most fragile. This certainly includes the unborn and newborn, but it also includes the mentally and physically disabled. This is why we oppose gene therapy when used for enhancement rather than for therapeutic purposes. Designer babies are but another way to discriminate against, and eventually to eliminate, those who fall outside society's norm.

God's Preferential Option for the Poor. The weakest members of society also include the poor who often lack resources and opportunity to speak for themselves. To believe in the sanctity of life includes supporting the poor by considering the economic impacts of our choices. As we have noted, genetically modified foods may accomplish a great deal in overcoming world hunger. However, with so much yet to learn, Christians need to be on the forefront of ensuring that policies and practices of governments and large agricultural corporations do not harm the poor.

The Golden Rule. Nearly everyone knows the Golden Rule, but not everyone realizes how helpful it is in the ethical issues surrounding genetics. The Golden Rule helps us realize that, just as we would want to be fed when hungry, we need to feed others. Assuming appropriate and necessary protections, this can include use of genetically modified foods. Those determining how genetic information is used and who has access to the benefits of genetic research should be guided by the Golden Rule.

The Golden Rule reminds us that just as we would not want someone experimenting on us without our permission, we should not allow research on others without appropriate consent. Babies born or unborn, others lacking the capacity to understand, and those vulnerable to coercion (for example, the incarcerated) deserve special protection with regard to medical testing. All of us, not just couples seeking in vitro fertilization, need to become informed of the ethical dimensions and implications of our decisions.

The Golden Rule should cause us to consider ecological consequences of our choices. Just as we would not want someone spoiling our land, we should be concerned about spoiling the land of others.

Genetically modified crops can have an ecological impact. They sometimes escape into the surrounding area, endangering native species. Insects that are not pests sometimes are negatively impacted. These are not just others' problems. The Golden Rule tells us we need to look out for others' well-being by advocating for appropriate safeguards. This does not mean we abandon research or use of genetically modified foods. Rather, we must move cautiously, seeking to do the least amount of harm while effecting the greatest good.

Culture of Death or Culture of Life?

The fundamental ethical issues surrounding genetics lie beneath the surface, beneath where we usually look. The real dangers lie not within the laboratory, but within the human heart. We now find ourselves in a culture of death which aborts infants because they are inconvenient, allows for euthanasia, treats human beings as valuable brains trapped in less valuable bodies, questions whether handicapped babies should live, and debates whether humans are better than other animals.

How did we get here? By choosing to value the individual more than the group, by assuming that morality is determined by democracy, by deciding self-actualization should be our highest goal. We got here by forgetting that every choice we make shapes our moral character. Instead of the culture of death, Christians should advocate for a culture of life. If death reigns when humans act autonomously, then life takes shape when we remember we are not God, commit ourselves to valuing what God values, and live by the Golden Rule.

Conclusion

We have landed on a new shore, ripe with both potential good and sinister dangers. Made in God's image, we cannot turn back to safer waters. Coming to this new land under the authority of the one who is greater, we are bound to obey him. We have a mandate to subdue this new territory, to understand it in its marvelous complexity, to play God. We cannot give way either to fear or to arrogance; we have both too much to gain and too much to lose.

Suggestions for Reflection and Action

1. Using commentaries and the works of respected theologians, research the Bible's teaching about what it means for humans to be made in the image of God. Genesis 1–2 is a good place to begin.

2. To become familiar with the issues surrounding genetics, read Francis S. Collins's, *The Language of God* or another good introduction.

3. Follow the news on genetic issues for one month. Get all the information you can from as many perspectives as you can. Evaluate the ethical issues involved in each issue against the two perspectives presented in this chapter: We are made in God's image; we are not God. For each, ask, "What could this news require me (or us) to do?"

For Further Reading

Collins, Francis S. *The Language of God: A Scientist Presents Evidence for Belief*. New York: Free Press, 2007.

Collins led the team that mapped the human genome. Along with much else, this work presents his conviction that DNA is the "language by which God spoke life into being" (p. 123). If you have wondered whether science and Christian faith are compatible, this book is for you.

May, William E. *Catholic Bioethics and the Gift of Human Life*, 2nd ed. Huntington, Ind.: Our Sunday Visitor Publishing Division, 2008.

May's book is an excellent introduction to many issues we could only briefly discuss in this chapter.

Visit the Web site http://www.who.int/foodsafety/publications/ biotech/20questions/en for current, useful information about genetically modified foods.

CHOICES BETWEEN LIFE AND DEATH

Christina T. Accornero and Susan Rouse

All the days ordained for me were written in your book before one of them came to be.

—Psalm 139:16

In recent years, social commentary on a variety of life-choice issues, both public and private, has become increasingly visible. Public opinion statements have evolved from being placed on simple bumper stickers and billboards viewed by a few driving by, to postings on YouTube, Facebook, Twitter, and other online venues seen by millions.

These statements made in public arenas might give the impression that scientific surveys have been conducted, opinion polls have gathered millions of responses, and everyone holds the particular opinion being sold. People research topics via the Internet, glean information from questionable Web sites, and become experts on the most difficult of issues.

> It is a poverty to decide that a child must die so that you may live as you wish.
> —Mother Teresa

This flawed information-gathering process now prevalent in society's decision-making can lead to disastrous consequences when life and death issues are involved. Because people have been led to believe the choices are easy, they often assume there really is only one acceptable choice or one intelligent view on any given ethical issue. A three-second glance at the YouTube list of abortion videos (accessed August 7, 2009) showed that one had been viewed well over three million times. While at one end of the life cycle, a popular bumper sticker proclaims, "It's not a choice; it's a baby," on the other end, Facebook lists over three hundred affinity groups for instruction guides, music, and places to go be euthanized.

As Christians, how are we to navigate this unending stream of information, opinion, and advice? This chapter explores the life and death choices at the beginning of life when pregnancy complications may arise, and at the end of life when suffering may seem to beg for a quick end. Examining these issues through the lens of Scripture, we seek both to understand what glorifies God and to discover more effective ways to discuss these issues in our larger social context.

Abortion

Human embryos and fetuses are just that: human. Few disagree. Rather, the most heated debates in bioethics today focus on the constitutional personhood of the unborn, not on their biological designation as human. Biologically, human embryos are destined to become human adults if exposed to the appropriate environment (a human womb). The ethical and legal issues, however, revolve around the question of when the developing human deserves constitutional rights and protection under the law. Clarifying this issue is critical at this point in history because many ethical issues hinge on whether or not a human embryo or fetus is a person with constitutional rights.

Roe v. Wade

What is the context of the abortion discussion today? Whether or not we believe abortion is ethical, the reality is that abortion is legal in the United States. It is legal because in 1973, the U.S. Supreme Court ruled in *Roe v. Wade* that a woman's autonomy, her right to privacy about decisions regarding her own body, is primary. The court ruled that since a woman possesses a privacy right to prevent pregnancy (for example, *Griswold v. Connecticut*), she also has the right to terminate a pregnancy. Given that this is the reality today in the United States, it is crucial that Christians explore and understand the issues.

The Wesleyan Church

In 1980, for the first time, *The Discipline of The Wesleyan Church* included a special direction for the denomination concerning abortion, and encouraged its members to become more informed.[1] *The Discipline* statement affirmed that the Church wanted to "recognize and preserve the inherent rights of life by opposing indiscriminately induced abortion for personal convenience or population control."[2] In 1984, the wording was changed, asking Wesleyans to take a decisive stand on the sanctity of human life. The 1984 statement stands in the current *Discipline* (2008): "The Wesleyan Church seeks to recognize and preserve the sanctity of human life from conception to natural death and, thus, is opposed to the use of induced abortion."[3] This stance suggests the bumper sticker may miss the point—it is both a choice and a baby! However, a deeper issue may emerge in cases of problem pregnancies: Which choice glorifies God?

Many argue that the early embryo is simply a collection of cells or a mass of tissue. However, a simple collection of cells cannot self-organize and develop into a human being. It is common to grow human cells in culture (in a petri dish in the laboratory), but these cells do not have the capacity to develop into an adult human. This feature is unique to the human embryo.

Biologically, from the moment of conception, an embryo is a self-organizing individual of the species *Homo sapiens*. Encoded in its forty-six chromosomes are all the instructions that will guide the development of that individual into a human person. No singular event happens after fertilization that determines its humanity.

Moreover, we all have a common beginning as human embryos. To deny an entire population of human beings justice and protection under the law simply because they are very young, they cannot speak, and they are dependent on others is not only inconsistent with Christ's teaching, but also is inconsistent with the U.S. Constitution (see the Fourteenth Amendment). A similar denial of justice and protection in any other context would be seen as a denial of civil rights.

In the very first chapter of Genesis, we are introduced to the *imago Dei*, the fact that humans are created in God's image (Gen. 1:26–27). The Genesis creation narratives together teach that God intended a special relationship with, and a special purpose for, humans. Moreover, Scripture seems to indicate that God sees each unborn as a person being formed in the womb with a purpose and a future in mind. Psalm 139 affirms that in no stage of development does God withhold involvement:

Oh yes, you shaped me first inside, then out; you formed me in my mother's womb. I thank you, High God—you're breathtaking! Body and soul, I am marvelously made! I worship in adoration—what a creation! You know me inside and out, you know every bone in my body; You know exactly how I was made, bit by bit, how I was sculpted from nothing into something. Like an open book, you watched me grow from conception to birth; all the stages of my life were spread out before you, the days of my life all prepared before I'd even lived one day. Your thoughts—how rare, how beautiful! God, I'll never comprehend them! (Ps. 139:13–17 MSG)

If God so directs human development from conception to natural death, it is reasonable to conclude that each human being, no matter the stage of development, bears God's image.

Scripture does not simply give us a holy perspective on the embryo as a human, bearing the image of God and known by God. It also instructs us to be fervent in protecting those who cannot protect themselves. One passage among many is Proverbs 31:8–9, "Speak up for those who cannot speak for themselves, for the rights of all who are destitute. Speak up and judge fairly; defend the rights of the poor and needy."

Autonomy

Autonomy is about choice. It is about fully capacitated adults having complete decision-making power over their own bodies. However, what the court did not speak to are the broader implications of our decisions. Very few personal medical decisions have no impact on others. We are autonomous adults, but we live in the context of our larger community and society. Scripture helps us understand that we are made to function as a body; we are parts connected to and influenced by one another's actions (see Rom. 12:3–8). When a woman makes the choice between remaining pregnant and having an abortion, her decision impacts the embryo or fetus, the father, and any extended family members. Her choice also impacts medical and health care professionals; lawyers may be involved; pastors or other counselors may be brought in—it is a personal choice, but with far wider implications than her own person. Exploring the psychological and sociological impact of abortion on all these is beyond the scope of this chapter, but we must address those most affected—the babies.

The choice may appear simple—choose life. However, almost every woman who has found herself facing a crisis pregnancy would beg to differ. Interestingly, most women participating in a study examining the psychology of the pro-choice mindset agreed that abortion is morally wrong in that it kills a baby, but stated, too, that it often is the only realistic choice for a woman experiencing an unwanted pregnancy.[4] By one estimate, nearly half of all women seeking abortion identify themselves as Protestants, and about 15 percent identify themselves as born-again Christians.[5] Clearly, many women in crisis feel as though they have only the one choice—abortion. As Christians living in community, how do we support women facing this difficult decision? How do we participate actively in the public debate that shapes law and public policy?

Many in churches around the world respond to these questions in tangible ways. Crisis pregnancy centers help women explore the possibility of carrying the baby to term. Christian adoption agencies work to find homes for babies born from crisis pregnancies. Many worthy and effective strategies exist for reducing the number of abortions. However, we also must be willing to speak in the public arena about alternatives to abortion. As we participate in the public debate, it is important for us to be focused and effective. Like Paul, we must understand our audience, and come to the public forum educated, ready to make a case that will effect change.

Education for Change

For years, most Christians arguing against abortion have focused on the sanctity of life and the personhood of the unborn child. The pro-choice movement, however, focuses on women's autonomy in exercising their reproductive rights. Ironically, the U.S. legal system seems to agree with both views. According to some laws, a fetus is human and does have rights. Thirty-six states currently have fetal homicide laws, criminalizing the death of (or the harm done to) an unborn child during commission of a violent act against a pregnant woman. These laws extend the definition of person to include a child in utero, at any stage of development, regardless of viability.[6] Ironically, these laws explicitly exclude abortion, but their very entry into our law codes bears witness to the fact that our legal system recognizes the inherent personhood of the developing human.

However, current abortion law focuses on the autonomy of the woman, and holds that her privacy rights trump the rights of the unborn human. For this reason, Paige Comstock Cunningham, one of the foremost Christian pro-life voices in the abortion debate, calls on Christians to change their focus. After more than thirty years of legal abortion in our country, we have a lot of information about how abortion affects women. Cunningham envisions a three-arc strategy that, when completed, will form an effective anti-abortion circle. The first arc is educating women about the ways abortion hurts women. The second arc consists in demonstrating that our society can live without abortion. The third arc, completing the circle, would establish in law that the U.S. Constitution protects unborn children and their mothers from abortion.

Cunningham sees our first priority as communicating to women that "abortion's seemingly quick resolution to [this] crisis is neither easy nor problem free; it creates more pain than it relieves. What's more, choosing birth is the truly courageous choice, one that resonates with [the woman's] inner values."[7] It becomes important to explore the negative consequences of abortion, as well.

Numerous studies have researched the psychological impact of abortion on women. These studies show that women who have had abortions are at greater risk for anxiety disorders, mood disorders, and substance abuse disorders, as compared with other women.[8] (Some of these studies compared women who had chosen abortion with women who had carried their first unwanted pregnancy to term.) The impact of abortion is not simply psychological, however. It is physical as well. Medical research has shown that women who have had an abortion have increased risk for a miscarriage in a subsequent pregnancy and increased risks of cervical cancer, ectopic pregnancies, infertility, endometriosis, and pelvic inflammatory disease.[9]

We must continue without addressing Cunningham's other two arcs. But it is clear even from this short discussion that abortion, as a quick fix to a difficult situation, often leaves women (and others) in an equally difficult place. The Wesleyan Church, with many others both Christian and non-Christian, is correct to affirm life. Could we also ask more energetically how we may help those facing the choice of whether to abort their pregnancies and those who have done so?

Euthanasia

Euthanasia, like abortion, is a complex matter. Likewise, euthanasia is also a quick fix for a suffering patient. However, as with abortion, we must not allow either autonomy or convenience to trump the sanctity of God-given life. Even when nearing the end of life, every human still carries and reflects the *imago Dei*; thus, end-of-life issues also merit our most thoughtful consideration.

The word *euthanasia* means different things to different people. Here, we define it as physician-assisted suicide or other such acts that induce death. Most would agree the intentionality of the act, whether by commission or omission, is the defining factor. Euthanasia also is referred to as mercy killing, an involuntary or voluntary end to someone's life for the purpose of ending suffering. When we come to the point of decision whether to prolong someone's life, however, we all hope the end will be without pain and suffering. This is where the pivotal question emerges: Who gets to decide when life ends?

Right-to-Die Laws

The modern movement to establish in law the right of a terminally ill person to choose the time and manner of death dates from the 1930s in Britain. In 1996, Australia's Northern Territory became the first jurisdiction to allow physician-assisted suicide. Also in 1996, Oregon became the first U.S. state to do so; early in 2006, the U.S. Supreme Court upheld the Oregon law. In November 2008, Washington State voters approved a measure similar to Oregon's. At the end of 2009, Montana's Supreme Court was considering an appeal of a lower court ruling allowing assisted suicide in Montana. Though most states now provide for persons to arrange for refusal of care as death approaches, to date, only Washington and Montana have followed Oregon's lead in allowing physician-assisted suicide.

The Wesleyan Church

In 1989, members of the Task Force on Public Morals and Social Concerns of The Wesleyan Church attempted to give Wesleyans guidance on issues related to death and dying. The Task Force recognized the complexities and benefits brought by advances in medical technology, while acknowledging that no one has all the right answers for individual and family decisions. No matter the medical advances, the Task Force concluded, and though we now are able to prolong life by many means, the day, hour, minute, and even second of death still is in God's hands.[10] We concur. We should note, too, that the Wesleyan *Discipline* does not address the issue of euthanasia.

Complexities

We continue to learn about how to make decisions affecting the quality of life as one approaches death. Many dedicated, wonderful health professionals provide care and dignity to those who are dying. The wide-spread system usually referred to as hospice care probably is the best-known example. Yet even here a difference of opinion arises. Most see hospice care as making possible a minimum of pain and suffering in the process of dying. Some, however, regard any and all intervention as an artificial push toward death. We must ask, then, how much is too much intervention in the dying process?

If God determines the moment of death, then any human act committed with the intention of ending life is considered euthanasia. Clearly, then, euthanasia includes administration of lethal doses of a drug, whether by mouth or by injection. Practically, euthanasia also includes an overdose of painkillers that would suppress the respiratory or the cardiovascular system. This definition, though, can become problematic when the patient is in a great deal of pain, because palliative care may include powerful narcotics that can and may inadvertently speed the dying process. The medical community has one central measure—a question—for separating palliative care from euthanasia: Is death intended? If the answer is no and the intent of the medication is to ease suffering, it is considered palliative care.[11]

A more difficult dilemma arises when the suffering patient requests death. Many do advocate euthanasia in this instance arguing that euthanasia: (1) provides a way to relieve extreme pain; (2) provides a way of relief when a person's quality of life is low; (3) frees up medical funds to help other people; and (4) provides individuals with freedom of choice.

Those on the other side of the debate who see a difference between prolonging life and intending death argue that euthanasia is simply an economically convenient solution that inappropriately empowers humans to end life. Their main arguments against euthanasia are: (1) euthanasia devalues human life; (2) euthanasia can become a means of health care cost containment; (3) physicians and other care providers should not be involved in directly causing death; and (4) a slippery slope effect occurs when euthanasia is legalized only for the terminally ill but, later, laws are changed to allow for other people to be euthanized non-voluntarily.

We know that God, in the person of Christ, suffered and died on a cross. Yet how do we enter into the pain and suffering of cancers, injuries, aging bodies, and other conditions caused by our residence in this less-than-perfect world? In the United States, we have access to amazing health care resources. Even the poorest of our poor are not subject to genocide, to the ravages of war, or to the myriad atrocities that plague many others. We are rich by comparison, so we must ask whether our riches drive our end-of-life decisions. Perhaps we have lived so comfortably we have lost sight of the spiritual values realized through aging and suffering. As Henri Nouwen reminds us, aging is both a promise and a treasure:

We believe that aging is the most common human experience which overarches the human community as a rainbow of promise. It is an experience so profoundly human that it breaks through the artificial boundaries between childhood and adulthood, and between adulthood and old age. It is so filled with promises that it can lead us to discover more and more of life's treasures. We believe that aging is not a reason for despair but a basis for hope, not a slow decaying but a gradual maturing, not a fate to be undergone but a chance to be embraced.

We therefore hope that those who are old, as well as those who care, will find each other in the common experience of aging, out of which healing and new life can come forth.[12]

Moreover, a biblical perspective on suffering at any age includes the view that suffering is a privilege (see Acts 14:22); that it draws us closer to Christ (see 2 Cor. 12:10; Rom. 8:17); and that when we rely on his strength to endure suffering, he is glorified in our weakness (see Phil. 1:20; 2 Cor. 12:9). God also promises to meet us in our suffering, to empower us to endure suffering with confidence (see Ps. 62:11–12), and to accomplish God's purposes through our pain (see Gen. 41:52; Isa. 64:8; Rom. 5:3–5; Heb. 12:7–8, 11). Sometimes those who must stand by and watch a loved one in pain have a harder time enduring the suffering than does the patient. In those difficult circumstances, God calls us to compassion (meaning "to suffer with") as we walk alongside those who suffer (Gal. 6:2), not to seek their quick end.

Conclusion

As Christians, would our view of abortion or euthanasia be different if we lived in India or Darfur? Christian friends in Somalia tell us that women in their country don't need to think about whether to have their babies or not even if a pregnancy is caused by rape. Though rape is evil, every child is a gift from God and is raised as such, as part of the family. Likewise, colleagues in Nepal tell us the family cares for those who are close to death, in the home if possible, and life ends when God is ready for it to end.

Thus, we come full circle, from life at the beginning to life at the end. Though our questions are still numerous, we see the beginning of life in the hands of a creative, life-sustaining God, and we see the end of life in those same hands. Therefore, we place our faith in this loving God who meets and carries us in our pain and suffering and who will carry us through to our final breath of life.

Suggestions for Reflection and Action

1. Begin to educate yourself about the dangers of abortion to women, both at the time of the abortion and in the years that follow; consider whether you can invite others in your church to join you. The Internet is a place to start, but be judicious in what information and conclusions you trust.

2. Consider volunteering with an organization that empowers women by providing alternative solutions and other services to those experiencing unwanted pregnancies. If no group is active in your community, explore whether your church may be able to begin one.

3. Educate yourself and help others in your church and community to educate themselves on the issues surrounding suffering, dying when it becomes an extended process, and death itself. Learn, for example, the difference between extending life and prolonging death.

4. Pray for wisdom to manage the pain and suffering of your own dying process and to participate appropriately in that of your loved ones. Discuss the dying process with family, and document your wishes through advanced directives.

For Further Reading

Colson, Charles W. and Nigel M. de S. Cameron, eds. *Human Dignity in the Biotech Century.* Downers Grove, Ill.: InterVarsity Press, 2004.

A collection of essays by Christian professionals working in fields touched by bioethics (natural scientists, physicians, attorneys, and clergy). Each essay is focused on a different bioethical issue and discusses the appropriate role(s) of Christians in the public discourse on these issues.

Kilner, John Frederic, and C. Ben Mitchell. *Does God Need Our Help?: Cloning, Assisted Suicide, & Other Challenges in Bioethics.* Tyndale House Publishers, 2003.

A short, non-technical work by two members of the Center for Bioethics and Human Dignity (a Christian bioethics organization), this book addresses today's major bioethical issues related to life and death from a biblical perspective.

Nouwen, Henri J. M., and Walter J. Gaffney. *Aging: The Fulfillment of Life.* New York: Doubleday, 1990.

A moving and inspirational collection of thoughts and photographs on what aging means and can mean to all of us.

Notes

1. *The Discipline of The Wesleyan Church 1980* (Marion, Ind.: Wesleyan Publishing House, 1980), 187.10.

2. Ibid.

3. *The Discipline of The Wesleyan Church 2008* (Indianapolis, Ind.: Wesleyan Publishing House, 2008), 410.11.

4. Paul Swope, "Abortion: A failure to Communicate," http://www.leaderu.com/ftissues/ft9804/articles/swope.html (accessed June 15, 2009).

5. Jim Kessler, "The Demographics of Abortion," http://content.thirdway.org/publications/17/Third_Way_Policy_Memo_-_TheDemographics_of_Abortion.pdf (accessed June 15, 2009).

6. National Conference of State Legislatures, "Fetal Homicide Laws," http://www.ncsl.org/default.aspx?tabid=14386 (accessed June 15, 2009).

7. Paige Comstock Cunningham, "Learning from Our Mistakes: The Prolife Cause and the New Bioethics" in *Human Dignity in the Biotech Century: A Christian Vision for Public Policy*, eds. Charles W. Colson and Nigel M. de S. Cameron (Downers Grove, Ill.: InterVarsity Press, 2004), 143–146.

8. "Psychological Risks: Traumatic Aftereffects of Abortion," http://www.unfairchoice.info/pdf/OnePageFactSheets/PsychologicalRisksSheet1.pdf (accessed June 15, 2009).

9. "Physical Risks: Life-Threatening Risks of Abortion," http://www.unfairchoice.info/pdf/OnePageFactSheets/PhysicalRisksSheet1.pdf (accessed June 15, 2009).

10. The Wesleyan Church Task Force on Public Morals and Social Concerns, "Position Paper on Issues Related to Death and Dying" (prepared by Lawrence W. Wilson, 1995), http://www.wesleyan.org/bgs/assets/downloads/Faith_Public_life/down.php?dfile=Death%20and%20Dying.pdf (accessed June 15, 2009).

11. John Frederic Kilner and C. Ben Mitchell, *Does God Need Our Help?: Cloning, Assisted Suicide, & Other Challenges in Bioethcis* (Carol Stream, Ill.: Tyndale House Publishers, 2003), 89–93.

12. Henri J. M. Nouwen and Walter J. Gaffney, *Aging: The Fulfillment of Life* (Doubleday, 1990), 19–20.

LIVING BY THE GOLDEN RULE

Jo Anne Lyon

*And whoever welcomes a little child like this in my name welcomes me.
But if anyone causes one of these little ones who believe in me to sin,
it would be better for him to have a large millstone hung around his
neck and to be drowned in the depths of the sea.*

—Matthew 18:5–6

My first time to Cambodia was July 1996. I thought I was prepared. I had read of its golden past. The ruins of the grand temples of Angkor Wat, almost nine centuries old, are now considered one of the ten wonders of the world. I had read diligently about Pol Pot and the atrocities of his rule during the late 1970s.

As I walked through the killing fields—there are more than three hundred—I could not imagine the evil that had so possessed a leader or how a people could follow so blindly. Within three years, at least two million people (some estimate more) were executed or died of overwork, disease, or other causes, and were thrown into pits—mass graves. These pits are now sunken,

> We need not buy exalted human worth at the expense of the rest of creation, but neither must we buy elevation of the moral status of other creatures at the expense of the extraordinary sacredness of human life.
>
> —David Gushee

grass-covered fields, but the victims' clothes, beginning to peek up through cracks in the ground, serve as a gruesome reminder of this horrendous genocide. Suddenly, the human pain they represented shouted louder in my subconscious than did the stark statistics.

Not only did the Khmer Rouge kill people, but they also delighted in torture, photographing its forms, means, and results. One picture—emblazoned in my memory—is of the minister of education in the previous government. As her torturers bored holes in her skull, even a tiny flinch or move to avoid or protest the pain only moved them to inflict greater suffering. In the photo, she is not flinching, but there is a lone tear on her left cheek.

Viewing all this as a civilized society, we want to shout, "This should not be!" Though the violent atrocities of Cambodia are in its past, such torture, brutality, and utter disregard for the value of human life continue in other parts of the world. Though it was adopted more than sixty years ago, the United Nations' Universal Declaration of Human Rights is often only partially implemented and sometimes completely ignored. I am grateful for the many human rights groups working to inform the global community of ongoing atrocities and to bring to prosecution the perpetrators of the many continuing human rights abuses.

The Misery of Human Trafficking

The killing fields were only the beginning of the story. At 10:00 a.m. on a Thursday, I came face to face with an incredible abuse for which I had not prepared myself in all my study of Cambodia. My host said, "Jo Anne, you need to go out with me to see this area. A few years ago, there were only three thousand young women and girls here; today they tell me there are some fifteen thousand. They are sold and also for sale."

When my host told me this, I recalled a brief article I had read in the *New York Times* on this subject a few weeks earlier. The article was horrifying, but I did not have any idea of the numbers. Now, as a few of us made our way down the dusty road with small wooden structures on either side, I could hardly believe my eyes. On the front of each of these rotting shacks was a small porch with young girls sitting on plastic chairs, all of them for sale! My heart could not accept what my eyes and brain were seeing.

Four of us took that walk that sunny morning. We never made it to the end as, truly, the end was not in sight. Finally, we stopped on a corner and, with incredibly burdened hearts, held hands and prayed. We were not a powerful group—three women (not young) and a male missionary in transition. There we simply opened our hearts, minds, and selves to God to use us in whatever way God chose, to value the highest of his creation, human beings made in his image.

I was not prepared for the streets of learning to which this prayer would lead. Since that day in 1996, *modern-day slavery* and *human trafficking* have become common terms to describe this horrific selling of flesh. In a document formally titled "Protocol to Prevent, Suppress, and Punish Trafficking in Persons, Especially Women and Children, Supplementing the United Nations Convention Against Transnational Organized Crime," the United Nations defined human trafficking as follows:

"Trafficking in persons" shall mean the recruitment, transportation, transfer, harboring, or receipt of persons, by means of the threat or use of force or other forms of coercion, of abduction, of fraud, of deception, of the abuse of power or of a position of vulnerability or of the giving or receiving of payments or benefits to achieve the consent of a person having control over another person, for the purpose of exploitation. Exploitation shall include, at a minimum, the exploitation of the prostitution of others or other forms of sexual exploitation, forced labor or services, slavery or practices similar to slavery, servitude or the removal of organs.[1]

This "Protocol" was signed on December 15, 2000, in Palermo, Italy, by heads of state and ministers of more than eighty United Nations member states, double the number required. With that, for the first time, trafficking in persons was defined in an international instrument, which could then serve, too, as model language and a common framework for nations adopting their own anti-trafficking laws. It also offered guidelines for implementation within individual nations, as definitions, prohibition, and enforcement should not be limited to transnational trafficking.

As in the nineteenth-century slave trade, human trafficking today is embedded in the economy of many nations. I find it disturbing that, as people look at this enormous evil, they tend to think it never will be abolished because of its economic power. It gives me pause to realize that with this attitude, our trust is in mammon (money) more than in God. As of this writing, revenue from human trafficking is less than that from trade in illegal drugs. Some, however, predict trafficking soon could surpass the illegal drug trade, noting that one can sell an ounce of cocaine only once, but a human being can be sold many times.

I had heard this statement often but did not grasp it in human terms until the director of the World Hope International Assessment Center in Cambodia e-mailed me with a follow-up report on a particular girl. The note said, "Today we received a thirteen-year-old girl in our center. She has never been to school, but speaks five languages. She was sold to a brothel in Cambodia when she was ten years old. Since Vietnamese was her first language, in Cambodia she learned Khmer. She then was sold to a brothel in Thailand where she learned to speak Thai. From there she was sold to a brothel in Malaysia and learned both Malay and Russian." In Malaysia, somehow, the International Office of Migration was able to free her and repatriate her to her home country of Cambodia. She was brought to the safety of the Assessment Center and there, finally, she has the opportunity for school and for psychosocial, medical, and spiritual healing.

The U.S. Department of State regularly updates its estimates of the number of persons held in slavery. Based on those estimates and the direction of the trend, it is accurate to say that, throughout the world, more than twenty-seven million persons are slaves today. The tentacles of the modern-day slave trade reach in many directions, so that for these people, their valuation, protection, and care as God's crown of creation are overwhelmed by the evil of greed.

Jesus' Care for Women and Children

The majority of today's twenty-seven million slaves are women and children. Most women and children were also near the bottom of the social order when Jesus walked the dusty roads of the Holy Land. It is critical, then, that we take note; these were two groups Jesus went out of his way to lift up, both in his words and actions, thus proclaiming them as made in his image.

Jesus even defined his kingdom through children, famously countermanding his disciples' attempts at crowd control to caress the children, hold them on his knee, and bless them emphatically, one by one (Matt. 19:13–15; Mark 9:36–37). In doing this, Jesus crashed through the cultural barriers of the day, as children were rarely even acknowledged publicly. One of the most powerful of Jesus' words he cast negatively (his warning that a millstone being hanged around one's neck and being thrown into the sea) was better than the fate of anyone offending one of these little ones (Matt. 18:6).

Certainly Jesus' respect, care, and love for women made the disciples very nervous. His discussion with the Samaritan woman stunned them. But even more shocking was that she was the first person to whom Jesus revealed himself as Messiah (John 4:26). Her response to his words and his love were not buried in an argument. She simply brought her entire town to Jesus.

Non-Trafficking Crimes Against Women and Children

Over the centuries, rape of women has been used as a weapon of war, too often unrecognized or unacknowledged. But as early as the eighth century B.C., Israel's prophets predicted widespread rape as one of the certain outcomes of the devastation foreign armies would visit upon Israel and Judah (Isa. 13:16; see also Lam. 5:11; Zech. 14:2). We could multiply examples across the centuries down to today.

Following World War II, however, military tribunals were established. Among their duties was prosecution for gender-based violence. The international law that guided their deliberations continued in effect some fifty years later as international criminal tribunals convened to prosecute cases of alleged rape from Yugoslavia and Rwanda. Instances of wartime abuse against women are now generally addressed according to the interrelated legal regimes of humanitarian and refugee law as well as human rights law.

In 1993, the Vienna World Conference on Human Rights adopted the "Declaration on the Elimination of Violence Against Women." This document defines and describes gender-based violence as something that results in or is likely to result in physical, sexual, or psychological harm or suffering to women, including threats, coercion, or arbitrary deprivation of liberty both in public and in private life.[2] While this declaration is not legally binding, it does bring onto the world stage the recognition of violence against women as a human rights issue.

Most recently, the United Nations Security Council unanimously passed a resolution on June 19, 2009, noting that "women and girls are particularly targeted by the use of sexual violence, including, as a tactic of war to humiliate, dominate, instill fear in, disperse and/or forcibly relocate civilian members of a community or ethnic group."[3] This resolution demands "immediate and complete cessation by all parties to armed conflict of all acts of sexual violence against civilians."[4]

Another way of abusing women and children is to use them as weapons of war. In a number of conflicts around the world, both boys and girls have been conscripted into rebel armies and coerced into doing dastardly deeds beyond their comprehension, often under the influence of drugs. Many of them die from the effects of the drugs and alcohol given them from an early age to make them compliant and capable of the heinous acts their captors force them to do. Some suicide bombings in recent decades fit this profile as well.

Environmental Disaster as Violence Against the Poor

Haiti

I arrived in Port-Au-Prince, Haiti, on a rainy night some thirteen years ago. The airport lights were dim. As I stepped outside, I heard my name and, "Follow me." I followed and found myself in water up to my ankles then soon on the run toward a waiting vehicle. As we drove through the streets, I heard constant screams and cries coupled with the sound of falling metal. The driver, rather casually and yet with concern, said, "Those are the people in Cité Soleil losing their homes. It happens every time we have a hard rain." To this day, those screams stay with me. I later visited Cité Soleil, and came to understand the incredible fragility of one of the largest and most vulnerable slums in the world. On that night drive, I had not been aware of the denuded mountains and the overworked soil.

India

A few months later, in north central India, I traveled through many cities with large, smoke-billowing, foreign-owned factories; I could barely breathe. One of the Indians in the car said sadly, "Yes, we have jobs, but we're dying earlier."

Most of us old enough to have heard about it then will remember the deaths of twenty-five hundred people from the first day of exposure to the insecticide methyl isocyanate in Bhopal, India, in November 1984. Defective storage tank valves at the Union Carbide plant allowed the invisible gas to escape in deadly concentrations. We are less familiar with the fact that this accident killed or permanently injured a total of two hundred thousand people, most of them residents of poor neighborhoods near the plant.

Zambia

I watched her thin hands clutch a small handmade shovel in her attempt to scratch the earth's hard-baked surface. The energy needed to dent the soil far surpassed the strength of this AIDS widow and mother of four young children. The hot day in Zambia carried several messages. The seasons are now unpredictable; the rains barely came during the rainy season that year, bringing on an early drought. This desperate mother, though, still was trying to plant a garden in borrowed space, hoping against hope to grow food for the youngsters for whom she was the only supply of nourishment. Her strength waned in the hot sun; her frail body, weakened by hunger, had little endurance. Yet, she also had no food or water for her family for that day, meaning she still faced several hours of walking and searching for something to eat.

Yes, Mutinta is made in the image of God. And, no, she is not aware of climate change. All she knows is the village elders have been saying something is changing, and they can't figure it out. People living close to the soil have interconnectedness with land and weather. But as one village elder said, "Things are very strange. We used to know exactly when to plant and almost the day the rains would start, but something very strange is going on that we have never experienced, nor did our ancestors."

Looking at the reality of climate change as it impacts Mutinta, her family, and her village, I could not help but think of the current debate about its causes. That day I realized the scientific and political wrangling could continue for decades. But Jesus calls me, and us, to a higher standard, "Love your neighbor as yourself" (Matt. 22:39, quoting Lev. 19:18).

How Do We Do That?

How do we love our global neighbors as we love ourselves? A good first step is to begin to see and experience how deep and broad and high is the importance God attaches to justice, mercy, and loving-kindness. Among hundreds of lines of Scripture intended to teach precisely this lesson, one specific passage stands out:

> So justice is far from us, and righteousness does not reach us. We look for light, but all is darkness; for brightness, but we walk in deep shadows. . . . Truth is nowhere to be found, and whoever shuns evil becomes a prey. The LORD looked and was displeased that there was no justice. He saw that there was no one, he was appalled that there was no one to intervene. (Isa. 59:9, 15–16)

Can we say we are as serious as God is about justice, mercy, and loving our neighbor? Since we cannot, repentance is our first order of business. But once God's conviction has drawn us to repentance, we may be tempted to despair. What can we possibly do? Human trafficking, global poverty, myriad oppressions, devastations of climate change, illiteracy—the list goes on—all are interwoven in today's complex world. It is almost impossible to separate one from the other. In our western minds, we tend to think in terms of silos, of each problem in isolation. However, one issue cannot be fixed without addressing the others. Here, we, the heirs of John Wesley's eighteenth-century leadership, may learn from the example of his and others' integrated approach—an approach that continues to influence our culture today.

Sunday School

Hannah Ball actually started the Sunday school movement later popularized by Robert Raikes. They began the Sunday schools as a literacy effort for poor children who had no other access to education, children who worked ten or more hours a day six days a week in factories and mines in the early decades of the Industrial Revolution. In Sunday schools, they taught these children to read, using the Bible as the primary text. The Sunday school movement marked the first step toward free education for all.

Poor Relief

John Wesley had a practical concern for the poor and contributed personally to their relief, along with raising funds from other sources. Wesley saw to it that, through his societies, clothing was distributed and food provided for the needy. Given his life-long interest in all things medical, it is no surprise the Methodists early on began to establish dispensaries to treat the sick. In London, Methodists turned one meeting room into a carding and spinning workshop.

British Christians opened a lending bank in 1746. Others made legal advice and aid available. Still others made sure widows and orphans were housed. Such Christian concern for the under-privileged led to the birth of the Benevolent (Strangers' Friend) Societies in 1787. These quickly established themselves as agencies of poor relief, bridging the gap until the state began to take on these responsibilities decades later. The Evangelical Revival made England aware of its social obligations.

Political Action

The British labor movement also is rooted in the Evangelical Revival. Wesley supported fair prices, a living wage, and healthy employment for all. At a time when coal was removed with hand tools, poor children as young as four and five years of age were put to work in the most remote areas of many mines; their small hands could reach where adults' could not. Evangelicals lobbied to rid England of this despicable oppression of children and succeeded.

Others of the Least and the Last

By long custom, the mentally ill had been hidden from public view. The efforts of awakened Christian conscience resulted in improved care and love for this segment of God's creation. Orphans were cared for. Religious training for young people led to the creation of the Young Men's Christian Association (YMCA) and the Young Women's Christian Association (YWCA). Other groups provided spiritual and social help for prostitutes, prisoners, the blind, and the deaf. Innovations such as the use of Braille resulted from this work.

Eighteenth-Century Prayer

One of the most important means to accomplishing these great social feats was prayer. Notably, a number of the wealthy were on the front line in the work of prayer. This was true particularly of the titled members of the Clapham church in the north of England, who spent three separate hours daily in prayer. Christians all over England united in prayer and connected their prayer to action, as well. The best-known example is their continual publicizing of the evils of slavery made tangible by organized boycotts of slave-produced goods.

Twenty-First Century Prayer

When I think of humanity as the crowning glory of our Lord's creation, I reflect on our habits of prayer, particularly our public and pastoral prayers. Rarely do I hear a public prayer in church for anyone or anything outside the local congregation. Most prayers and prayer requests are for the sick among their own number. What would happen if we should begin earnestly to pray for an end to human trafficking? Would God answer a prayer for the increase of micro-finance programs that give the poor dignity and new life? Dare we pray for the violent wars to cease in which women and children not only are forced to wage war, but are themselves used as weapons of war?

The Unavoidable Summons

Annie Dillard, in her outrageous style, reminds us of the power of God available to the Christian:

On the whole, I do not find Christians, outside of the catacombs, sufficiently sensible of conditions. Does anyone have the foggiest idea what sort of power we so blithely invoke? Or, as I suspect, does no one believe a word of it? The churches are children playing on the floor with their chemistry sets, mixing up a batch of TNT to kill a Sunday morning. It is madness to wear ladies' straw hats and velvet hats to church; we should all be wearing crash helmets. Ushers should issue life preservers and signal flares; they should lash us to our pews. For the sleeping god may wake and take offense, or the waking god may draw us out to where we can never return.[5]

"Draw us out to where we can never return!" It is in this space we will learn the width and depth of loving our neighbor as ourselves. Do we dare dream of the transformation both of people and of the planet?

Suggestions for Reflection and Action

1. Research the following issues in and for your community: What is the status and frequency of domestic violence in your community? Does your community have shelters for women or their affected children? What services exist for the batterer? What are the roots of violence against women?

2. Seriously consider accepting this challenge: What would happen if your church prayed for thirty days about the pain and suffering in your community and in the world—prayed both privately and publicly? If you do accept this as a call of God upon your church, record the doors of service that open and the results as your congregation walks through them.

3. Challenge yourself and your small group or your congregation to repeat the Golden Rule every morning for thirty days and track the opportunities God opens to live it out concretely. Make opportunities to share both the difficulties and the joys you encounter.

For Further Reading

Cannon, Mae Elise. *Social Justice Handbook: Small Steps for a Better World*. Downers Grove, Ill.: InterVarsity Press, 2009.

This is a comprehensive resource of models of ministry, as well as practical exercises in taking action to make a difference in the issues we have discussed. Echoing Micah, it is a handbook for living justly, loving mercy, and walking humbly with our God.

Kristof, Nicholas D., and Sheryl WuDunn. *Half the Sky: Turning Oppression into Opportunity for Women Worldwide*. New York: Knopf, 2009.

Kristof and WuDunn are married to each other; both write for *The New York Times*; both are Pulitzer Prize winners. Their research and stories of personal encounter bring the subtitle to life. The book also shows their respect for people of faith and is, in fact, a subtle call for more people of faith to respond to these issues.

Mertus, Julie A. *War's Offensive on Women: The Humanitarian Challenge in Bosnia, Kosovo, and Afghanistan*. Bloomfield, Conn.: Kumarian Press, 2000.

As the title suggests, Mertus traces the history of the numerous kinds of abuses of women perpetrated by men at war. This work is a good introduction to war's enormous effects beyond the battlefield.

Sider, Ronald, Philip Olson, and Heidi Unruh. *Churches That Make a Difference: Reaching Your Community with Good News and Good Works*. Grand Rapids, Mich.: Baker Books, 2002.

This book is based on an analysis of fifteen urban churches working in holistic ministry. The principles presented here may be used in any setting.

Notes

1. "Protocol to Prevent, Suppress, and Punish Trafficking in Persons, Especially Women and Children, Supplementing the United Nations Convention against Transnational Organized Crime," http://www.uncjin.org/Documents/Conventions/dcatoc/final_documents_2/convention_traff_eng.pdf (accessed September 5, 2009).

2. United Nations General Assembly, "Declaration on the Elimination of Violence against Women," http://www.un.org/documents/ga/res/48/a48r104.htm (accessed September 5, 2009).

3. United Nations High Commissioner of Human Rights, "Rape, Weapon of War," http://www.ohchr.org/EN/NewsEvents/Pages/RapeWeaponWar.aspx (accessed September 5, 2009).

4. Ibid.

5. Annie Dillard, *Teaching a Stone to Talk: Expeditions and Encounters* (New York: Harper Perennial, 1982), 52–53.

PART 3

CARE OF THE ENVIRONMENT

LAND AND WATER CONSERVATION

Travis H. Nation and Kenneth D. Dill

*The earth is the LORD's, and everything in it, the world, and all
who live in it; for he founded it upon the seas and
established it upon the waters.*

—Psalm 24:1–2

The Importance of Land and Water

Land and water are important to God. They play a major role in both the opening and closing chapters of the Bible. In Genesis 1:1, the beginning is marked by God's creation of the heavens and the earth. The creation account reveals that God crafted a system that included a planet with atmosphere, land, seas, and other waters. God's presence was there. This planet was the perfect place to sustain plant, animal, and human life.

The last chapter of the last book of the Bible draws us to a

> Just as we look back on previous times with incredulity and wonder how people, especially believers, could have not only condoned but succored the slave trade and slavery, so in later years I think subsequent generations, who will live consciously with the reality that the earth is not a limitless larder, will find it difficult to understand how we could have described ourselves so uncritically as "consumers."
>
> —James Jones

new heaven and a new earth. Revelation 22 tells us there is a river of the water of life. Its origin is the throne of God. The river cuts its way through the center of the city. This river of life waters the many specimens (not just one!) of the tree of life standing on either bank of the river. Each month a crop of fruit is produced and the leaves of the tree will heal the nations. God is fully present there. It is the perfect place.

Land and water are important to us. It can be argued they are Earth's most precious natural resources. Other than breathable air, no other resources are as vital to our existence as these. The case for the practical importance of land and water is not hard to make. Land-based green plants capable of photosynthesis provide a vital link between life's primary energy source, the sun, and organisms unable to use that energy directly. These plants are the foundation of the world's food supply.

Additionally, many of life's most important chemical reactions take place in solutions where water is the solvent. Most organisms, including humans, cannot function without water. Dependable land and water sources have historically determined the locations of human population centers. Stated most simply, life as we know it could not exist without adequate land and water.

Land and Water Conservation: Whose Priority?

Despite the unquestionable worth of land and water as natural resources, the dawn of a new millennium finds land and water conservation low on the list of priorities for most individuals, people of faith among them. Regrettably, the twentieth-century Christian church was not known for environmental activism and resource conservation. Whether due to emphasis on end-times theology, fear of liberal political agendas, or merely widespread indifference, modern Christianity has been reluctant to embrace environmental concerns.

However, it is not difficult to make the case that land and water conservation efforts are not only biological necessities, they are biblical mandates. We may begin with the observation that, in biblical times, the people of Israel valued both land and water very highly. They had to. In the edge-of-the-desert geography of the Holy Land, land suitable for agriculture is very limited. From the beginning of Israel's monarchy onward, arable land increasingly was controlled by the wealthy and powerful elite.

Compounding the problem, water was also scarce. In the heart of Israel's territory, the Central Highlands west of the Jordan River, none of the streams flowed year-round. Olive trees, grapevines, and fig trees survived the summer dry season relatively well. However, barley and wheat, sown in late fall and harvested from late spring to early summer, both required sufficient rainfall at the right times or the crop would fail. Relatively minor variations from the standard yearly patterns, early or late, virtually assured famine in the affected regions the following winter.

Every settlement away from the few natural springs (the "living water" of several important biblical texts) required hewing numerous cisterns from the limestone bedrock and lining each with a waterproof plaster made from slaked lime. Cistern water sustained people, livestock, and kitchen gardens through the summer dry season. However, cisterns were easily contaminated. The few sewer systems that did exist consisted of channels, covered with flagstones, running down the middle of the main streets. Seepage of human and livestock waste into cisterns fostered the rapid increase of disease-causing microorganisms. Though they did not know these existed, people knew the addition of wine to water in a ratio of about one-fourth to one-third usually kept the water from making them sick.

The scarcity of arable land and potable water elevated them to the status of precious commodities. Their physical value in biblical times is reflected by their importance in Scripture. We see it in the garden of Eden, the flood of Noah, in the journey to the promised land, and in the use of agrarian features in hymns, psalms, and parables. The very sacraments we celebrate are derived from these two essentials of the earth. The Lord's Supper uses two products of the land: wheat and grapes. The physical water of baptism symbolizes the inner cleansing and renewal of the Christ-follower. In addition to their life-giving physical properties, land and water clearly manifest spiritual symbolism as well.

For those in the developed world, technology has made high quality food in the grocery store and clean water at the tap little more than afterthoughts. These presumptions have led to complacency which serves as the foundation for the modern lack of concern. While land and water seem abundant and are considered renewable resources, they do not exist in infinite quantities. Unfortunately, few appreciate the relative scarcity of usable land and water in the world today. Current data and worldwide trends call us to reconsider the state of these precious resources.

Worldwide Land Availability and Use

Approximately one-third to one-half of Earth's land surface has been modified by humans. Historically, most of this modification has been for agriculture to feed the world's ever-growing population. This situation produces an ironic dilemma, if not on a global scale, at least on a regional scale: more people, less land; less land, less food; less food, fewer people.

Granted, technology has led to a green revolution. Fertilizers and pesticides, mechanization and irrigation, and high-yielding hybrid crop plants have made possible dramatic increases in agricultural production possible. However, this scientific advancement has not come without a price in the form of grave environmental consequences. Fertilizers and pesticides have increased demand for fossil fuels (mainly oil) and introduced abnormally high levels of chemical residues into soil and water resources. Overall, the diversity of food crops has decreased with the majority of humans living on just twelve species of plants and approximately three-quarters of all human food coming from just eight species.[1] Moreover, large-scale agriculture often destroys natural habitats, further decreasing local, regional, and global plant diversity.

More recently, the problem of land use has been compounded by the choice of crops agricultural land is actually used to grow. A significant portion is used to grow feed for livestock to supply the developed world's demand for meat. Feeding plant material to animals which will be used for human consumption adds an additional energy-wasting link to an already naturally inefficient food chain.

In addition, the use of biofuel crops such as soybeans, sugar cane, corn, and other grains to produce ethanol or bio-diesel further reduces the amount of land used for food production. Not so incidentally, if use of biofuels does indeed lessen the developed and developing world's dependence on fossil fuels, the large-scale diversion of major crops to this purpose also results in much higher food prices, through simple supply-and-demand economics. Though all of us are affected, these increases hit (the poor) hardest.

If space allowed, we would discuss a number of other land use problems. Poor urban planning—where to put what—often leads to traffic and pollution problems and to regular flooding of streets, homes, businesses, and vacant land without adequate storm drainage. Infrastructure replacement, repair, and expansion—streets, water, and sewer lines—tend to be postponed because they are very expensive. Postponement often intensifies the problems of pollution, flooding, erosion, and other concerns. All these directly affect the land under the feet of those who live there. As with Hurricane Katrina, for example, the effects don't always stop at the city limits.

Worldwide Water Availability and Use

While land use problems are pressing, the increasing scarcity of usable water is an even greater concern. Roughly three-quarters of Earth's surface is covered with water, yet less than 3 percent of it is fresh water. Furthermore, almost 90 percent of that small freshwater fraction is locked in glaciers and icecaps or is otherwise too far underground to be useful. The remaining minute fraction—the total of accessible fresh water—is only two one-hundredths of 1 percent (0.02 percent) of all the water on Earth.[2] Relatively speaking, usable fresh water is rare!

In addition to its scarcity, there are other varied and widespread problems with water. In the last one hundred years, water demands for industrial purposes, agriculture, and personal needs have increased more than twice as fast as population growth. Worldwide, the average annual per capita use of water for all purposes is approximately one hundred sixty thousand gallons.[3] Some countries have the resources to support that demand, while others are forced to barter with other nations for water or utilize expensive desalination methods.

Virtually every inhabited continent has an ongoing environmental saga involving water depletion. The Aral Sea, located between Kazakhstan and Uzbekistan in central Asia, has lost approximately 80 percent of its water over the last thirty years. In central Africa, Lake Chad is a fraction of its original size. In the United States, the Ogallala Aquifer, a large underground reservoir stretching from Texas to South Dakota and supplying the nation's agricultural heartland with irrigation water, has decreased rapidly over the last fifty years. In virtually all cases, water depletion of such magnitudes is due to inefficient large-scale crop irrigation. Research, education, and training in more efficient watering methods are reasonable strategies for addressing these present predicaments and preventing future ones.

In the developed world, water shortages are inconveniences: We can't water our lawns; we can't clean our streets; we can't wash our cars. In the developing world, water shortages are survival issues: lakes dry up; crops fail; children die from starvation, dehydration, and waterborne diseases. The United Nations estimates that approximately one billion people lack access to safe drinking water, and within fifteen years, roughly half the world's population will live in water-stressed areas.[4] The present humanitarian crisis in the Darfur region of Sudan is possibly caused, and definitely compounded, by water shortage.[5]

The argument for land and water conservation is strong. In an effort to prevent the promotion of an alarmist mindset, we have intentionally avoided using words like *catastrophe*, *disaster*, and *emergency* here because, in actuality, most of the world is not yet at the crisis point. However, we must get serious about asking the questions: What should be our response toward places where land or water shortages are a present crisis? What can we do now to prevent land and water availability from becoming a crisis in other places?

Of course, the answers to these and related questions are multi-faceted. They may differ for each individual or group reading this chapter. Some may choose to buy locally grown produce; some may decrease their consumption of meat; some may install low-flow toilets; some may practice water-saving landscaping around their homes; some may embark on numerous other approaches that facilitate responsible use of land and water resources. A thorough discussion of conservation strategies is beyond the scope of this chapter. However, the old axiom, "Think globally, act locally," could be an appropriate universal philosophy from which to begin.

An Attitude of Conservation: Motivating Factors

Perhaps an additional question for the twenty-first century church is: Should our response regarding land and water conservation be any different than the rest of the world's response? For people of faith, the answer should be an unequivocal, "Yes!" When it comes to conservation of a limited resource, one of three forces usually motivates individuals to change their behavior: (1) physical necessity, (2) economic pressure, or (3) unselfish obligation (such as, conscience).

For example, consider what it would take to get a hypothetical community to change its attitude toward water conservation from one of indifference to one of concern. If water resources suddenly became so scarce that individual households were limited to fifty gallons of water per day, we would expect attitudes and consequently behaviors concerning appropriate uses of water to change because of physical necessity. Likewise, in a less critical scenario, if the price of municipal water increased in proportion to household use, families would probably watch their use of water more carefully and cut out wasteful practices due to the increased economic pressure. In fact, some proponents of free market economics might suggest that conservation will happen only when it becomes economically necessary. Ultimately, however, from a Christian perspective, the most desirable type of conservation would involve individuals voluntarily limiting water use due to a sense of responsibility toward fellow citizens and future generations, acting out of unselfish obligation before the diminishing resource level became critical.

One final additional factor capable of motivating the church toward conservation is worth considering: positive peer influence. Even when there are no physical, economic, or ethical reasons to get involved in environmental issues, individuals sometimes can be persuaded toward that end by members of their own social group. If influential people exhibit conservation-friendly behaviors, those behaviors will become more popular and will be emulated. Regardless of the source of motivation, the positive consequences of this type of influence will make it worthwhile.

Conservation: The Default Christian Response

God intended for us to care for the earth, not neglect it. And God certainly did not intend our attitudes and actions to contribute to its destruction. Genesis 1:28 gives us our job description: "God blessed them and said to them, 'Be fruitful and increase in number; fill the earth and subdue it. Rule over the fish of the sea and the birds of the air and over every living creature that moves on the ground.'"

Implicit in this instruction is the old adage, "Do no harm." In the obtuseness of our now-sinful nature, though, we have tended to mis-interpret it. Instead of becoming caring gardeners, we often have taken on the role of abusing consumers. God directed us to be stew-ards of creation; we are not to exploit or deplete our resources. When we do, we create stress on other parts of the earth's ecosystems. Frequently, we do not notice this trauma until it has reached a crisis point. Then, we have to focus our attention and assets on correcting that predicament. Marshalling our energy for the crisis takes attention away from another area. Moreover, in our efforts to fix one problem, we often create another one. If instead of crisis responders we became imitators of God, we could model responsible long-term management, seeking to balance ours and the earth's needs and to experience all cre-ation as a reflection of God's goodness.

In his sermon "The General Deliverance," John Wesley addressed particularly the ethical treatment of animals and in general the Christian's response to brute creatures and creation. This was forward thinking for Wesley's time. He opposed hunting as sport, contests that caused pain to animals, and other cruel treatment of animals. Wesley understood that all creatures are a part of the creation. He knew that God, upon completing this creation, had approved it and pronounced all of it good. Wesley reasoned that because God has shown mercy to us, we also are to show mercy to the rest of God's creation. Because God equipped humanity with the power of reason, we must use our knowledge and reason to pay forward the mercy extended to us. Wesley admonished his hearers, "So much more let all those who are of a nobler turn of mind assert the distinguishing dignity of their nature."[6]

The concept of self-discipline is important to this topic. It is always tempting to do what interests us or what will be to our advantage. It is part of our sinful nature to be selfish. The first murder recorded in the Bible occurred because Cain wanted the approval Abel received. Fueled by his selfish desires, Cain rejected God's counsel, deceived his brother with an invitation to join him in the field, and killed him.

We may not have attacked and killed our brother with our hands, but we find it very easy to build expensive houses on the shores of barrier islands. Wanting the ocean view and easy access to the beach front, we do not think about what we will do with our wastewater. We do not concern ourselves with how our structures will change the wind patterns that shift sand away from the vegetation protecting the shore from erosion. Consider another example: Profit motivates some to clear-cut large areas of rain forest and sell the timber to international corporations. Run-off water from the newly naked land flooding the village below is not their concern.

When we become followers of Christ, we are asked to place our passions and appetites under the authority of the Holy Spirit. Christ models self-discipline for us. Our environmental ethic should reflect this attitude of self-discipline. Just because we can does not mean we should.

As people of faith, unselfish obligation should be the default response in matters of conservation. After all, Jesus repeatedly spoke to the necessity of putting the needs of others before our own and treating our neighbors as ourselves. As followers of Christ, our efforts toward conservation cannot wait until it is all but necessary or compulsory. An attitude of conservation within the Church should be a natural outflow of our commitment to Christ, not something we have to be coerced to do.

Additional Benefits of a Christian Environmental Ethic

While utility certainly is important to conservation, we should note that our concern for the earth's resources is not based exclusively on what is useful to us. God created the earth for enjoyment, placing our first parents in a specially-planted garden. It contained all they needed, not only for their physical well-being, but also for their emotional and psychological well-being. Still today, we identify with this first human home so completely that its very name symbolizes a perfect paradise where all our needs are met: Eden.

At the end of each creation day, God declared that day's work good. At the end of the creation process, God declared it all very good. God's creation was without sin or the destructive force of evil. Everything worked in concert with everything else. Genesis 2:9 reads, "And the LORD God made all kinds of trees grow out of the ground—trees that were pleasing to the eye and good for food." The Hebrew word translated "pleasing" also can be rendered "delightful." Creation is delightful to the eye. It is pleasing to the senses.

What is pleasing to your eyes? What delights your senses? A crisp autumn morning with leaves crunching under your feet? The crashing of waves and the feel of salty air on your face? Watching wind-driven snowflakes come to rest clinging to bare branches? The songs of migrating warblers? We all celebrate some form of creation that causes us to pause and wonder. God uses these moments to remind us there is such a thing as beauty. In a fallen world with so many reminders of the scars of sin, we need to do all we can to preserve the beautiful, the pleasing, the delightful. They are reminders of God's original creation and of the glory that will be on display when God renews his creation with a new heaven and a new earth.

People of faith acting in one accord can serve as a powerful example in matters of environmental concern and resource conservation. Would God expect anything less of us?

Suggestions for Reflection and Action

1. Using a concordance, look up references to land or water in the Bible. If you check them all, this will be a long project, so we advise beginning with a single book—for example, Genesis or John. Where did people find water? What attitudes to land and water do you see recorded? How did people use land and water? Are land and water used as metaphors or as representative of other blessings from God?

2. Check the Web site of World Hope International, www.worldhope.org (or another international aid and self-help organization). Investigate what you, your family, or your local church could do to help people dealing with the land and water problems discussed in this chapter. Decide on one action to get you started, then plan it out and do it.

For Further Reading

Au Sable Institute of Environmental Studies. http://www.ausable.org. Promotes Christian environmental stewardship through academic and community programs, retreats, and conferences.

Austin, Richard Cartwright. *Baptized into Wilderness: A Christian Perspective on John Muir*. Louisville: John Knox Press, 1987.

Part of a multi-volume set focusing on environmental theology, this book provides glimpses into the life of John Muir with implications for Christian involvement in environmental efforts.

Bratton, Susan. *Christianity, Wilderness, and Wildlife*. Scranton, Penn.: University of Scranton Press, 2009.

An examination of the tradition of Christian wilderness spirituality from Noah's and Moses' experiences in the Old Testament to Celtic monasteries and the Franciscan order. Bratton traces a long history of divine encounters in biblical literature such as visions, providential protection, spiritual guidance, and calls to leadership—all of which highlight the importance of nature in Christian thought.

Environmental Protection Agency, Water Sense Program. http://www.epa.gov/watersense/index.html.

A partnership program sponsored by the U.S. EPA seeking to protect the future of our nation's water supply by promoting water efficiency and enhancing the market for water-efficient products, programs, and practices.

The Sierra Club. "Faith in Action: Communities of Faith Bring Hope for the Planet." http://www.sierraclub.org/ej/partnerships/faith/default.aspx.

Report on the environmental engagement of communities of faith; highlights one exceptional faith-based environmental initiative from each of the fifty states, the District of Columbia, and Puerto Rico. Demonstrates the diversity of spiritually motivated grassroots efforts to protect the planet.

Notes

1. Food and Agriculture Organization of the United Nation, "Facts and Figures about Plant Genetic Resources," http://www.fao.org/english/newsroom/action/facts_ag_treaty.htm (accessed March 8, 2010).

2. William P. Cunningham and Mary Cunningham, *Environmental Science: A Global Concern* (Boston: McGraw-Hill, 2007), 377–381.

3. David Seckler, Randolph Barker, and Upali Amarasinghe, "Water Scarcity in the 21st Century," *International Journal of Water Resources Development* 15 (1999): 29–42.

4. United Nations International Decade of Action Water for Life, 2005–2015 "Fact Sheet on Water and Sanitation," http://www.un.org/waterforlifedecade/factsheet.html.

5. Lydia Polgreen, "How Much Is Ecology to Blame for Darfur Crisis?," *New York Times*, July 22, 2007, http://www.globalpolicy.org/component/content/article/206/39774.html.

6. John Wesley, "The General Deliverance," *The Works of John Wesley*, vol. VI (Kansas City, Mo.: Beacon Hill Press, 1979), 252.

CHRISTIAN STEWARDSHIP OF NATURAL RESOURCES

Richard L. Daake and D. Darek Jarmola

The LORD owns the earth and all it contains,
the world and all who live in it.

—Psalm 24:1 NET

An important challenge facing modern Christians is how to reconcile theological arguments of God's provisions for creation with ethical questions concerning apparently wasteful uses of those resources. As believers, we need to learn and accept the scope of our responsibilities toward God's creation. Questions of how we use natural resources—for instance, how we get around—are theologically significant.

One promising approach is the concept of eco-stewardship, addressing at least three major environmental issues: (1) drilling and mining, (2) wasteful use of

> The Possessor of heaven and earth . . . placed you here, not as a proprietor, but a steward.
>
> —John Wesley

natural resources, and (3) sustainability both of resources and of the environment. Here, we will address directly the issue of wastefulness and only indirectly the other two areas. For the purposes of this discussion, we will define *eco-stewardship* as intentional, biblically based, human responses to and responsibility for the provisions and resources we find in God's creation.

In this chapter, we hope to help contemporary, mostly affluent Christians understand biblical principles of eco-stewardship. We will track briefly the history of Christian responses to the environment and identify attitudes toward wastefulness found in American Christianity. Finally, we will urge an ethos of eco-stewardship that more faithfully reflects biblical, Christian teachings and practices.

Eco-Stewardship and the Bible

One regrettable misunderstanding of Christian stewardship comes from misreading a key element of the Genesis creation story. Many Christians have interpreted God's intention to give humans "dominion" (Gen. 1:26 KJV) as having kingly rule over God's creation. This view of stewardship tends to skew our attitudes and actions toward expedient, self-serving control. We become conditioned to view God's creation as only a necessary element supporting our existence. This in turn leads to a presumption that humans are both separate from and above the rest of God's creation. This is not a biblical view.

Within the creation narrative, Genesis 2:15 is the strongest corrective to this mistaken reading, reporting that God placed the first human in the garden to "work [or even serve] it and to take care of it." We are caretakers of resources that are not our own, responsible to keep the creation at least as valuable and as fertile as when we came to it.

Long before humanity faced resource limits and environmental problems of waste, the Bible suggested a simple, yet profound, solution. In its context, the *bal tashchit* (bahl tahsh-CHEET, "ch" as in loch) principle of Deuteronomy 20:19–20, "do not waste or destroy," prohibits destruction of fruit trees during the siege of a city. Jewish tradition has elaborated on this to frame four guiding principles: (1) use only as much of a resource as you need; (2) do not needlessly destroy any resource; (3) do not use a resource of greater value if something of lesser value is available; and (4) do not use a resource in ways contrary to its design, as this increases the likelihood of its being damaged or destroyed. A modern equivalent of these principles is the slogan, "Reduce, reuse, recycle."

In the episode of the feeding of five thousand, Jesus instructed his disciples, "Gather the pieces that are left over. Let nothing be wasted" (John 6:12). Jesus' action reinforces the *bal tashchit* principle and provides several theological insights: (1) God is not wasteful; (2) God expects us to be prudent and frugal in our use of resources; (3) wasteful, wanton use of resources is contrary to a biblical way of life; (4) all resources are precious, not to be thrown away without thought for tomorrow or for the needs of others; and (5) we are to handle excess resources wisely and carefully even if they are waste.

A further theological dimension of eco-stewardship is the biblical expectation that, as all nature was impacted by the fall, so all nature ultimately will be redeemed. If we respect nature and its resources only because it is the right thing to do, we assume an ethical stance that is neither uniquely nor entirely Christian. Christian faith ascribes value to natural resources because God is the source of their existence. Thus, respecting creation and natural resources is one important way of honoring God and living Christianly.

It is important to understand our task of caring for creation as a responsibility to God. Alexander Shapiro writes, "We are enjoined [by God] to look well at our relationship to the environment that surrounds us, to view it as an aspect of God's holiness, and to treat it with deference and respect."[1] Shapiro went on to suggest that a biblical worldview demands we squelch personal and cultural greed and view our stewardship in two broad contexts: creation as belonging to God, and history as requiring us to leave a proper inheritance of resources and godly attitudes for the generations that will follow us. In sum, the biblical principle of eco-stewardship demands our thoughtful care of all God has entrusted to us.

Eco-Stewardship and the Church

Until recently, an intentional, consistent concern for stewardship of natural resources has been largely absent from Christian discourse about biblical theology and ethics.

To the Reformation

The history of the early Church is one of persecution and survival under the Roman Empire. Even after Constantine's establishment of the Christian faith, the idea of stewardship of God's resources is addressed only indirectly as when, in his *Confessions*, Saint Augustine wrote that "wastefulness is a parody of generosity." Augustine equated wastefulness with vices such as pride, anger, aggression, sensual indulgence, willful ignorance, sloth, self-pampering, stinginess, envy, cowardice, and melancholy. Clearly, then, Augustine saw stewardship as the positive and practical expression of one's faith. Yet, such stewardship expressions often were inward-looking (duty toward my possessions), rather than focused outward on the broader environment.

The medieval church, too, was defined mostly by its struggle for survival. This time, the physical threat came from relentless Muslim invasions. In addition, Islam introduced continuing threats to Christian theology and orthodoxy. Within Christian Europe itself, poverty and vice brought about by political and social abuses of feudalism sorely tested ethics and morality. The realities of prolonged military conflict, plague, famine, and disease were all too common throughout the Middle Ages, destroying or otherwise wasting precious resources, both human and natural.

These adversities, however, led medieval Christians to practice stewardship, both via charity and by the necessary simplifying of their dependence on material goods. A strong consensus arose that it was wrong to waste any of God's gifts. For example, if food was left over from meals, it was reused, given to the poor, or used as animal feed. However, just as with the early church, there is no evidence of interest in providing theological reasons for this necessary stewardship.

The Reformation to the Present

During the Reformation, the Anabaptists were the first to articulate the importance of creation care within the framework of Christian discipleship. Anabaptist theology, with its dependence on agro-theological communities, taught that planting of seeds for the kingdom of God required responsible action toward the physical environment. While marginalized in its own time, such a view of eco-stewardship began to be accepted by mainline Christianity toward the end of the Enlightenment and the beginning of the Industrial Revolution. As a result, the Anabaptist claim that salvation brings environmental issues under the lordship of Jesus Christ is now a central theme in the conversation about Christian stewardship, generally.

John Wesley did not address the issue of wastefulness or eco-stewardship directly either in his sermons or his other teachings. In several of his sermons he defined stewardship as "the present state of man." Most of Wesley's teaching on stewardship deals with issues of money and business. However, Wesley did see believers as God's stewards who must act according to God's will and design. He came close to our eco-stewardship principle in his preaching and teaching that God has entrusted us with worldly goods to create opportunities for doing good with the ultimate purpose of advancing the happiness of our fellow human beings. Wesley also taught that good stewards must improve upon the resources left in their care by the creator.

Today's Church and Eco-Stewardship

Contemporary Christianity has either adopted or transformed many secular ecological strategies. By 1993, the Evangelical Environmental Network (EEN) was formed to "educate, inspire, and mobilize Christians in their effort to care for God's creation."[2] One of the most transformative acts of this movement has been to clarify the terms of Christian stewardship to mean dominion on the earth (as opposed to over the earth) for the sake of the environment and living beings. In practical terms, this means that wise and effective use of resources, not wastefulness, must be the driving force behind Christian eco-stewardship. At the very least, senseless waste of natural resources is both unbiblical and disrespectful. At most, waste is an outright sinful act, because it is contrary to the covenant God established with humanity.

A major driver of Christian environmentalism today is the Roman Catholic Church. Pope John Paul II wrote that the key to Christian stewardship is "not only to limit the damage which has already been done, and apply remedies, but especially to find approaches to development which are in harmony with respect and protection for the natural environment."[3] These pronouncements and actions (Pope John Paul II and the EEN) begin to introduce biblical, theological principles into what has been predominantly the moral dimension of an expanding environmental crisis. The concept of eco-stewardship gives preservation its needed theological dimension.

Eco-Stewardship in North America

Most followers of Christ agree that wastefulness is bad practice and should be stopped. Most take the theological step of acknowledging that the earth and all its inhabitants are the Lord's (Ps. 24:1). Translating the ethical and theological principles into changes in personal behavior, new cultural mores, and changes in the law where necessary is what requires hard work and long-term commitment. Except for tiny changes, these are usually more difficult than expected.

For individual Christians and local congregations, this may begin by comprehending and establishing the goal of being good eco-stewards of the natural resources provided by the creator. Recycling materials, reducing demands on limited material and energy resources, and using resources more effectively and efficiently are all part of this equation. In some cases, we may realize that where complex technologies are involved, solutions to one problem may create other unintended, negative consequences for the environment. We need to think globally and Christianly about such issues and be ready to revise our opinions as technologies change while maintaining our commitment to the underlying biblical principles of eco-stewardship.

What Is a Resource?

Illustrating the importance of culture and cultural attitudes is the very fact that what counts as a resource for a given society, in its time and place, is also culturally determined. This depends not only on whether a material or object is readily available, but also on the presence or absence of relevant technologies. Fresh water for drinking, cooking, and bathing is valued by all people; all would perish without it. Until recently, enough water—and clean enough water—has been available to most people without excessive cost and effort. Conversely, platinum is very rare, but was not accorded high value until the recent discovery of its effectiveness as a catalyst for newly important chemical reactions, such as reducing toxic emissions from internal combustion engines.

Quartz crystals of high purity have always been valued for their beauty. However, not until the need for precise timing circuits in watches, computers, and a host of other modern instruments were these crystals seen as a natural resource of industrial value. Most societies highly value gold, but a few have ascribed it almost no value. The high value accorded the diamond, both as a precious stone and as an industrial workhorse, is relatively recent in most western societies. Long-playing vinyl records have gone from being a home entertainment necessity to an oddity or a collectable.

We could multiply examples, of course, but the point is that what counts as a valued resource depends on cultural needs and available technologies. As technologies change, so will the processes of identifying resources and the resultant generation of waste products.

What Is Waste?

Some examples of wastefulness have become so commonplace in everyday life we may fail to recognize them. The ubiquitous throwaway plastic of bottles, bags, and shrink-wrap packaging both degrades the environment and threatens the survival of individual small animals that run afoul of it. Yet many of us seldom notice it. In our use and handling of plastic, as with many other materials, we may be unaware that a particular practice is wasteful, or we simply may accept it as a necessary result of the way we live.

According to the United States Environmental Protection Agency (EPA), in 2008, the United States generated approximately two hundred fifty million tons of trash or about 4.6 pounds per person per day. The EPA also reported that less than 33 percent of waste is recycled; the rest is either incinerated or buried in landfills.[4] Of course, both incineration as waste treatment and trash disposal in landfills can (and often do) pose serious hazards to the environment in general and to human beings and animals in particular. The practice of eco-stewardship here (reduce, reuse, recycle) would lower the demand on natural resources and potentially diminish environmental, economic, and public health problems.

How Can We Change Our Habits?

In our fast-paced, North American society, we purchase our soft drinks and even much of our water in plastic bottles. We buy disposable diapers and thousands of other items wrapped in layers of plastic and cardboard. Then we carry them all to our fuel-guzzling SUVs or crossovers in plastic bags! Most of this we do from habit, driven by our desire for convenience.

Now, good arguments other than convenience can be made for some of these cultural "needs"; improved health due to improved sanitation would be one. Still, concern for the wise use of resources, including the energy needed to make these products, together with the problems of their disposal, will prompt a more thoughtful discussion within the Christian community. A biblical concern for creation care can and should provide one important context for such discussion.

A Recycling Success Story

The modern wood products industry provides many good examples of transforming trash into treasure. Scrap pieces of logs, beams, boards, plywood, and, of course, giant heaps of sawdust used to be regular features of sawmills large and small across the continent. They were nuisance waste materials, fire hazards, and in a few locations, even a hazard to safe travel. Now virtually every particle including the sawdust is converted into particle board, fiber board, innumerable paper products, and a host of other useful products and materials.

For those committed to creation care, such examples of sustainable recycling of a spectrum of related wastes into useful products generally elicit positive emotional responses and do deserve thoughtful attention. That reducing the amount of waste materials to be discarded also can be good business practice would seem only to be a bonus. It may be tempting to assume that anything that can be recycled should be recycled.

A Recycling/Disposal Necessity

However, because recycling processes themselves also place demands on material, energy, and financial resources, decisions about whether to recycle certain materials can be very complex and difficult. Recycling is not always the right creation care answer, but recycling decisions should not be made solely on the basis of economics either. Sometimes recycling is financially less than a break-even transaction, but we should recycle, anyway, for creation care reasons.

One example is lead storage batteries, still widely used in vehicles and other applications. Unfortunately, lead is a very hazardous material, and it will make its way gradually into groundwater if these batteries are disposed of in landfills. Most developed nations, including the United States and Canada, now regulate the disposal of worn-out batteries to prevent lead contamination. Local congregations and other Christian groups can practice eco-stewardship by reminding their members of the proper storage and disposal of lead and other common hazardous waste materials, such as paint and many lawn care chemicals. Assistance with disposal, offered to those less able to do it on their own, will benefit the land, the water, and every living thing dependent on them.

A Decision to Recycle or Not

Unlike lead storage batteries, many modern materials can be disposed of safely in landfills. Glass, for example, poses little hazard to the environment or to drinking water when disposed of properly. It may be safe to decide solely on economic grounds whether to recycle it. In the case of glass, the economic decision, too, may be for recycling, because recycled glass often brings at least a small profit.

However, the plastics we spoke of earlier pose a wider range of disposal issues. Some plastics are easily decomposed by a combination of sunlight and microorganisms. Others persist much longer.

Those plastics, such as polyethylene, that occupy the largest portions of our landfills generally do not produce significant amounts of hazardous products as they decompose. The main problem with these plastics is their sheer volume. Burning them would reduce their impact on landfills, but burning most plastics poses more serious environmental problems (air pollution, release of greenhouse gases, and so forth) than does placing them in landfills.

Should we recycle plastic, then? For some plastics, the cost of recycling is excessive and hard to justify. Others can be recycled economically into wood replacement products and other composite materials for use in construction.

The value of plastics goes well beyond convenience in packaging, though convenience is a value. Plastics keep water and food safe for consumption. Plastic piping keeps contamination out of groundwater. In thousands of ways, plastics make our lives safer and more convenient. Plastics even save lives, both in routine and in spectacular ways. As eco-stewards we must learn to weigh the thousands of applications of the various plastics against the challenges of their disposal as wastes, or the cost of their conversion into other usable materials when we are finished with them as plastics.

We have barely begun, but we must bring our discussion of Christian eco-stewardship to a close.

Conclusion

The Christian faith is deeply rooted in belief in God as creator, redeemer, and sustainer. As a material reality, this world and those creatures that inhabit it, including human beings, were made for God's pleasure and glory. God intends not only that humans care for the earth, but that we use the material resources of God's provision to honor and serve him and to help in the accomplishment of God's purposes here.

We are neither to worship the creation nor abuse it; we are to be good stewards of it. Still, it is not enough merely to preserve the good earth any more than it was acceptable in Jesus' parable of the talents for the unprofitable servant to preserve what his master had given him to care for (see Matt. 25:14–30). God expects his children to invest what has been placed in our hands to accomplish his purposes. We are to avoid both active wastefulness based on selfish motives and inaction arising from fear. Fortunately, there is ample solid middle ground.

To be a good steward of the earth means to use its natural resources wisely in ways pleasing to God; to appreciate its beauty; to use its bounty in ways that encourage all human beings to see and appreciate its blessings as gifts from God; and to participate with God in bringing forth, even from natural resources, the potential for good that God intends.

As Christians, we should be suitable examples in all aspects of our lives. Respecting God's creation is one significant way of showing others that we love God and honor him with our actions. To behave well toward what God has made—the world God enjoys and declared to be very good—is itself a form of worship. Living as good eco-stewards is an expression of our love for and faith in God, an important expression that brings God glory.

Suggestions for Reflection and Action

1. Consider: Does an abundance of resources tend to produce attitudes of wastefulness and carelessness toward the environment? If your answer is yes, how should the Christian church in North America respond to these attitudes? How should Jesus' command for his followers to be salt and light in our culture affect our own attitudes toward waste and wastefulness? How can or should we impact our culture to be less wasteful of natural and man-made resources

and more respectful of the earth and our environment? How can or should our daily living patterns change?

2. Consider: Does the so-called prosperity gospel promote poor stewardship and wastefulness among Christians? What examples come to mind? If your answer is yes, what should you do about it in your life?

3. A case study, perhaps for a group: conversion of starch from grains, such as corn to ethyl alcohol has been technically possible for a long time. It provides a renewable source of fuel for internal combustion engines and the solid by-products from this process are used as animal feed. However, production of alcohol from grain crops reduces the food available for human consumption. Two questions (at least) arise: (1) Is this wasteful or not? and (2) is this a wise use of resources in light of widespread human hunger?

For Further Reading

DeWitt, Calvin B. "Preparing the Way for Action." *Perspectives on Science and Christian Faith* 46 (June 1994): 80–89.

DeWitt identifies three biblical principles foundational to environmental stewardship. He also discusses legitimate environmental concerns, showing how they easily can become unbalanced, leading either to inaction or to overreaction.

Funk, Ken. "Thinking Critically and Christianly About Technology." *Perspectives on Science and Christian Faith* 59 (September 2007): 201–211.

Funk, a Christian engineer, discusses wise and ethical uses of technologies based on biblical principles of human responsibility to God, our fellow humans, and the rest of creation.

Jenkins, Willis J. *Ecologies of Grace: Environmental Ethics and Christian Theology*. New York: Oxford University Press, 2008.

Jenkins deals with the relationship of salvation to environmental issues. He argues that Christians ought to be at the forefront of creation care debate and action, first for theological reasons and only secondly for pragmatic reasons.

Snyder, Howard A. "Salvation Means Creation Healed: Creation, Cross, Kingdom and Mission." *The Asbury Journal* 62:1 (Spring 2007): 9–47.

Snyder is in the top rank of Wesleyan scholars writing on creation and creation care issues. Here, he sets both creation care and human responsibility in a biblical and distinctively Wesleyan context.

Notes

1. Alexander Shapiro, "Sharing our Planet," *Chemtech* (August 1984): 456–457.

2. Evangelical Environment Network, www.creationcare.org/about.php (accessed September 1, 2009).

3. Pope John Paul II, Catholic World News, "Respect for environment is sign of hope, Pope says," http://www.catholicculture.org/news/features/index.cfm?recnum=8939&repos=4&subrepos=1&searchid=591721 (accessed March 15, 2010).

4. Environmental Protection Agency, "Municipal Solid Waste Generation, Recycling, and Disposal in the United States: Facts and Figures for 2008," http://www.epa.gov/epawaste/nonhaz/municipal/pubs/msw2008rpt.pdf (accessed March 15, 2010).

EVERY LIVING CREATURE

Ronald R. Crawford and Joseph Coleson

Don't muzzle an ox while it is threshing.
—Deuteronomy 25:4 MSG

Chickens caged for life so we can have inexpensive eggs for breakfast. Surgery without anesthesia on animal research subjects. Dog fights under cover of darkness. Exotic pets turning on their owners. We've all heard a hundred such stories of our horrible treatment of animals. Some occur without notice 365 days of the year. Others are sensational enough to grab public attention for a day or two when the stories break.

> Faith in Jesus Christ can and will lead us . . . to the broader concern for the well-being of the birds in our backyards, the fish in our rivers, and every living creature on the face of the earth.
> —John Wesley

It doesn't have to be this way.

Jasmine

Jasmine. The name evokes in me (Crawford) a feeling of gentleness and delicate acceptance. Jasmine—Jazzie, as I lovingly called her—was my cat. I chose her but not at first. I grew up on a farm where dogs were productive workers, and cats seemed mostly a hissing, scratching nuisance. Oh, they kept mice and rats under control in the barn, but they reproduced at an astonishing rate and were always overrunning the place themselves.

But Jazzie! My wife and I first encountered this calico kitten as a scrawny starveling emerging from a drainage ditch and yowling at us in desperation. My wife said, "No, Ron. We have six dogs now; we do not need more mouths to feed." But this kitten stayed with us and continued yowling. She moved between my legs and stayed there as we walked the last ten yards to our front door. I stopped, looked down, and said, "Honey, I can't stand it. I have to, and I don't know why!"

From then on, Jasmine was a member of the family. When I came home from teaching at the university, she would come running. Jasmine even talked! To get attention or acknowledge our arrival in the house, she would vocalize a *hellooooow* that definitely was not a meow.

In Persian, Jasmine means delicate flower, and she was that. She never was strong and, despite all our attempts to help her in a last illness, she died at the age of nine. I was crushed the day I watched her breathe her last. My relationship with Jasmine represents, I believe, a little of how we are to view our relationships with animals generally.

The Web of Life

God designed life on earth with a remarkable set of commonalities. All living things share properties and processes that are more alike than different, especially on the level of the cell. Beyond that, although humans are more than animal, we assuredly are part of the animal kingdom. To understand this more clearly and to see some of the implications for our treatment of animals, we will revisit briefly a few points discussed in the first section of this book—but from a slightly different perspective.

Creation's Interdependence

God designed this creation in an orderly progression—from non-living material, to plant life, then to animal life—and the living organisms are nurtured by materials provided from the non-living environment in intricately designed biological processes of transfer. From single-celled algae to multi-cellular complexes, plants use water to absorb minerals into their cells and transport the various dissolved nutrients to photosynthetic centers. Within these centers, molecular structures capture the energy of sunlight and fire the chemistry of life, transforming and storing it in the form of edible plant starch and releasing oxygen for atmospheric support of life. Once the plant life of the third creation day was established (Gen. 1:11–12), an ordered and balanced nutritive source existed for the various forms of animal life—the birds and sea creatures of the fifth creation day and the land creatures of the sixth creation day (Gen 1:20–31).

Altogether, God designed the earth's biological life in and for reciprocity, a balanced harmony of life that, in its totality, biologists call the food web. Atmospheric oxygen, continuously augmented by plant processes; atmospheric carbon dioxide, continuously augmented by animal (including human) processes; the sum of all biochemical processes in plants and animals; the waste of the various excretions incident to all life; and finally, the recycling of organic materials into further soil enrichment at the death of any organism not consumed by another—all these demonstrate the mutual relationship and dependency of the plant and animal kingdoms, of life as God designed it for this earth.

The Water of Life: More than We May Know

One further marvel: Both plant and animal life depends on the special design of the water molecule. Water consists of one oxygen atom and two hydrogen atoms, geometrically constructed in a V-shape. This is accomplished by the more electronegative oxygen atom nucleus pulling the electron from each of the two weaker hydrogen atoms and exposing the positive proton nuclei of the two hydrogen atoms. This creates a slightly negative oxygen end and a slightly positive hydrogen end to the molecule, giving water molecules a polarity that accounts for their ability to dissolve other polar substances.

All life consists mostly of water. The internal environment of plants must have water for all the chemical processes leading to the production of plant starches. In turn, animal life, to assimilate these nutrients and convert them into usable energy sources for the chemical fires of cellular life in animals, must be mostly water to produce an environment for use of the nutrients. All the materials necessary for life are water soluble and, therefore, are transportable in the watery environments within every living organism.

Our biology itself demonstrates God's intentional interconnection of all life upon this earth. God's physical, material creation is good. In important ways—though of course not in every way—we are related to all earthly life. We should respect and take appropriate care for life as one acknowledgement that, indeed, we are "fearfully and wonderfully made" (Ps. 139:14).

Guidance from Scripture

An important principle of biblical interpretation is that we do not establish a teaching by a single occurrence or reference. However, both the Old and New Testaments contain scores of references to human treatment of animals. We will mention only a few here, with minimal comment, to demonstrate that our first parents' rebellion did not end our responsibility toward the animals God has placed under human care and protection.

Creation Day Six: Our Kith and Kin

We highlight here a single point from the creation story of Genesis 1–2. In two places, Genesis 1 designates the animal creation as "living creatures," Hebrew *nephesh hiyah* (vv. 24, 30). Genesis 2:7 then says the *'adam* also "became a living being" (*nephesh hiyah*). Moreover, Genesis 2:7 reports God "formed the ['*adam*] from the dust [or clay] of the ground ['*adamah*]," and Genesis 2:19 says God "had formed out of the ground ['*adamah*]" all the larger beasts of the field and birds of the air God brought to the *'adam*.

With respect to our physical existence, then, we are part and parcel of the animal kingdom. The other creatures of the earth really are our physical kith and kin. Saint Francis of Assisi recognized this, and called them "little brothers." Another way to say this is to acknowledge that though humans are more than animal, we are not less. If we take seriously this physical kinship, God's instruction to care for and protect the rest of the animal creation only formalizes what we would have expected to do, naturally, before the first rebellion.

The Rest of the Pentateuch

Some would argue the single most important Old Testament record is the Decalogue, the Ten Commandments, recorded in Exodus 20 and Deuteronomy 5. In Exodus, the fourth commandment includes "your animals" in the instruction for Sabbath rest (20:10). Deuteronomy 5:14 is more specific: "neither . . . your ox, your donkey, or any of your animals" is to work on the Sabbath. Animals need and deserve a Sabbath rest, along with the humans for whom they work.

Deuteronomy 25:4 directs the farmer not to muzzle the ox that pulls the threshing sledge around the threshing floor, separating heads of grain from the stalk. Allow the animal to snatch occasional mouthfuls while it is working. The ox, too, deserves its incidental rewards.

Deuteronomy 22 contains several animal welfare provisions. The Israelite farmer was to collect a stray ox, sheep, or donkey and keep it until its owner could retrieve it (vv. 1–3). He was to help get an ox or a donkey, fallen under its load, to its feet again. Exodus 23:4–5 extends this obligation to the animals even of one's enemy. The young in a bird's nest could be taken, but the mother must be left (Deut. 22:6–7). This was both ecological wisdom and a humanitarian act.

The prohibition against hitching an ox and a donkey together for plowing (Deut. 22:10) could spare either. If the weaker donkey could not pull its weight, the ox bore the whole burden. If the stronger ox hung back, the donkey had to overcome not only the resistance of the plow, but also the weight of the ox on the yoke.

We can only mention here God's joy in and care for the animal creation as described in Psalm 104; Job 38–41; and throughout the book of Proverbs. A special case is the celebration of nature, including both flora and fauna, in the Song of Solomon. The lovers' joy in physical creation expresses worship, profoundly and truly.

Jesus' Words and Other New Testament Commentary

In one discussion with legalists, Jesus is famous for proposing the case of the ox fallen into a ditch; even his opponents would do the humane thing and work to get it out on the Sabbath (Luke 14:5). On another occasion, Jesus remarked that they would not leave an ox or a donkey thirsty in the stall all day because the day happened to be the Sabbath (Luke 13:15); they would lead it out to water.

In his Sermon on the Mount, Jesus taught that God provides food for the birds of the air and clothes in splendor even the grass of the field (see Matt. 6:25–30). On another occasion, he affirmed that not even a sparrow can fall to the ground, unless if God allows it (Matt. 10:29).

Important also is Paul's statement that the whole creation groans, waiting for its redemption along with ours (Rom. 8:19–21). God cares for all of his good creation.

Balaam's Long-Suffering Donkey

We've saved the most unusual example for last. It's the fascinating story of the patient beast we thoughtlessly call Balaam's ass. This story records the great steward himself sending an agent—we usually call them angels—to defend the humble donkey from her master's injustice (see Num. 22). Balak, king of Moab, sent for Balaam, a prophet of Yahweh. Ultimately, God said Balaam could go with Balak's men, but Balaam was to do and say only what God told him.

Imagine Balaam's surprise, embarrassment, and anger, when for no reason he could see, his donkey turned off the road into a field. He beat her with his staff to turn her back. A bit later, at a narrow place between two vineyards, she pressed close to a stone wall, crushing Balaam's foot against it. He beat her again. Finally, the angel of God, whom Balaam had not yet seen, stood in a narrow place where turning aside was impossible, and the donkey lay down under Balaam. A third time, he angrily beat her with his staff (vv. 21–27).

Balaam's anger may have moved him past it by this time, but we certainly are surprised to read that the donkey protested, asking, "What have I done?" (22:28). Balaam's response is telling, "You have made a fool of me! If I had a sword in my hand, I would kill you right now" (v. 29). The donkey reasoned with her unreasonable master, "Have I been in the habit of doing this to you?" (v. 30). Balaam was forced to answer, "No" (v. 30).

Only then did God allow Balaam to see the divine messenger from whose ire his faithful donkey had saved him. Imagine being called out by an angel for your savage treatment of the patient beast who just had saved your life—not once, but three times! The angel told

Balaam, "If she had not turned away, I would certainly have killed *you* by now, but I would have spared *her*" (v. 33, emphasis added).

Balaam had set his heart on the riches promised by Balak and cared nothing for his donkey. Even before her temporary power of speech, however, she was trying to speak to Balaam by her actions. Only his donkey's faithfulness and God's mercy saved Balaam's life that day. Moreover, the contrast between God's attentive concern for the donkey and Balaam's callous disregard should sober us all.

This story requires us to ask whether we really are different from Balaam. Are we different in the ways we look to acquire wealth? Are we complicit in mistreatment of animals, getting them to produce as much and as fast as they can, so we can gain greater wealth more quickly? The apostle Peter warned his readers of people like Balaam and even mentioned Balaam's donkey in his homily (2 Pet. 2:15–16).

Peter understood that our present human condition strongly tempts us to invest with inferior value other things that are not ourselves. We operate in our own wisdom and give rein to our greed for quick gain, snubbing the relationship with God that enables us to function as good stewards. This marring of the image of God in us must be fixed, restored, and redeemed to bring us back into harmony with creation. Without this regeneration, our care of the animals in our charge too often degenerates into abuse.

The Rise of Agribusiness

For much of human history, what we have called—in our own agrarian past—family farms were almost entirely self-sufficient. Farmers produced most of the food and feed consumed by their families and livestock and most other necessities as well. Cattle, sheep, and beasts of burden often were given individual names by their owners or by the children of the family. Such an environment fostered benevolent care for animals. The owner who neglected or abused his animals was regarded with disdain as both cruel and unwise.

High-volume production for maximum profits began in the eighteenth century with crops such as bananas, rubber, tea, coffee, and sugar cane in what we would now call developing countries. When these methods are transferred to North American production of corn, wheat, soybeans, and a few other major crops, they are sometimes given the (usually negative) label of mono-culture farming. Applied to the production of livestock—mainly cattle, hogs, and poultry—it sometimes is called factory farming.

This kind of farming in North America, increasingly corporate rather than family, has developed and accelerated since about the mid-twentieth century. Its impetus stems largely from the need to keep up with the expenses of operating larger agribusiness farms. A few dairy cattle, a few beef cattle, and a few hogs cannot support a family operation in today's agricultural economy. Livestock operations must be designed to bring in as much income as possible from as little investment as possible. Given this necessity of scale, the temptation is always present to ignore or cut corners with the best interests of the stock, short of inducing illness, injury, or death.

This is not to say farmers today are heartless fiends—far from it. Even on strictly practical grounds, they have to maintain the health and well-being of their livestock to get the best market prices possible, and most farmers are far better than merely practical. The alert and informed consumer is, however, an extra protection against the abuse of livestock destined for the market.

Affluence

In recent decades, the vast increase in disposable income in the developed world gives the average person choices never before possible and, along with them, a brand new range of ethical decisions.

Eating Meat

Since the development of agriculture, that is, of growing grains as the staple foods, most people have eaten meat only on family, religious, political, and a few other holidays and special occasions. Of course, hunting and fishing cultures are exceptions, but in agricultural societies, regular meat-eaters were almost exclusively the political and priestly aristocracy and the very few, mostly merchants, who made it to the ranks of the super rich.

In today's global economy, millions of people, including most North Americans, expect to eat meat at least once a day. Christians who choose to eat meat should consider where their meat comes from as one responsibility under our mandate to care for the earth's animals God placed under our care and protection. Is our beef made affordable by inhumane feedlot and slaughtering operations or by indiscriminate use of antibiotics to keep the cattle from obvious, observable illness? Do our eggs or chickens come from caged layers or broilers that never see sunlight and are never allowed even minimal exercise? Does our pork come from factory operations where the animals are under constant stress?

A few animal foods seem to be produced mostly by methods that include intentional cruelty. For example, in the United States, calves raised for veal usually are kept in cramped conditions from birth, allowed neither light nor movement. This is done to increase the tenderness and delicacy of their meat. Can we justify this for the sake of a more pleasurable dining experience, especially as humane methods of raising veal calves are known and available to all producers?

Exotic Pets

Chimpanzees, lions, tigers, and constrictor snakes all have made the news in recent years for injuring or killing their owners or others, including small children. We would not argue that no one should ever keep an exotic pet. However, Christians considering acquiring one should examine all the possibilities from beginning to end: Where does this animal come from? How was it acquired? Do I have the space it really needs, or will it always live in cramped conditions? Can I be sure it will never escape to harm itself, other animals, or humans? Can I be sure no one can reach it to harm it or to release it without my knowledge? Have I really considered every possibility, however far-fetched and ridiculous it may seem? Can I afford the expense of feeding and caring for it, not just for the first year, but for its expected lifespan? Can I justify the expense of providing for this pet, rather than using the money to sponsor a missionary or provide a scholarship to a seminary or Christian university? Could I gain the fulfillment I am looking for by volunteering at a zoo?

Food for Life

Some Christians, considering these issues, have become vegetarian or vegan in their food choices. Some point to the argument that if no one ate meat, the grain crops each year could feed the earth's entire population, rather than allowing more than one billion people to live in perpetual malnourishment. From Genesis 1:29–30, it is possible to conclude meat was not part of God's original intention for human diet. Abstaining from eating meat as a way of lessening animal suffering can be the principled decision of a committed Christian.

On the other side of the issue, Genesis 9:3 records God's granting Noah and his family permission to eat meat following the flood. The Pentateuch prescribed for the Israelites which animals are clean and which are unclean, but Israel never considered Gentiles bound by those instructions. In the early church, the same question was resolved in favor of Gentile freedom not to worry whether animals are clean or unclean according to the Torah (Acts 15:28–29).

Peter's vision recorded in Acts 10:9–16 is also relevant. Peter was praying on the roof of a home at noon and became hungry. He fell into a trance and saw a large sheet let down from heaven by its four corners, containing a variety of creatures, including reptiles and birds. A voice instructed Peter to kill and eat. Peter replied he never had eaten anything impure or unclean. The voice responded with a command, "Do not call anything impure that God has made clean" (v. 15). In Peter's vision, this happened three times, and the sheet was taken up again.

What was God telling Peter, and us, through this vision? (For this discussion, we are setting aside the main point of the vision: Peter's preparation for an invitation to visit a Gentile home.) One point of the vision is that all things are available for our use, but not for our abuse. Meat eating is not condemned. Determining what is right to eat is not a matter of legalism in our interpretation of Scripture. We may consider dietary laws, lists of healthy foods, and other helps, but we should do so in light of all the knowledge we can acquire to help us come to rational decisions about healthy eating and healthy living.

John Wesley on Our Treatment of Animals

In several of his sermons, John Wesley urged the humane treatment of animals working the mines, farms, and city streets of eighteenth-century Britain. He based his teaching on the fact that animals are God's creatures; God knows them and cares how we treat them.

Wesley went even further, however. He believed and taught that not only humans, but animals also, will be part of the resurrection life. For Wesley, simple justice demanded that animals now forced to lead "horrid" lives will partake of the joys awaiting all creation when released from its bondage to decay, as promised by Paul (Rom. 8:19–22). If animals will be our companions in the restoration of all things, should we not treat them as well as we can in the here and now?

Suggestions for Reflection and Action

1. Study for yourself the Scripture passages referenced and discussed in this chapter to arrive at a firm biblical foundation for your own understanding of how God intends humans to treat animals. Note where you agree with the authors, where you disagree, and why. This may be a good study project for a small group in your church or community.

2. Research the origins of meat and other animal foods available to you. Could you make changes in your purchasing habits that would be a vote for the humane raising and processing of these foods?

3. Consider the question: Should affluent Christians reduce our consumption of meat, especially of expensive luxury foods, to have more resources to give for the alleviation of world hunger and other worthy goals? If your study and discussion leads you and your small group to an affirmative answer, how could you begin? Craft a plan that would get you started and help you achieve the goal of giving tangible help to those who need it—in your community, across the globe, or both.

For Further Reading

Webb, Stephen. *Good Eating* (The Christian Practice of Everyday Life series). Grand Rapids, Mich.: Baker Press, 2001.

This book could be described as an introduction to the Christian ethics of eating. You may or may not agree with Webb's positions, but you will find your thinking and understanding clarified.

Wesley, John. Sermon LVI, "God's Approbation of His Works"; Sermon LX, "The General Deliverance"; Sermon LXIV, "The New Creation"; Sermon LXVII, "On Divine Providence"; in *The Works of John Wesley*, Kansas City, Mo.: Beacon Hill Press, 1979.

You may not agree with all Wesley's statements and opinions in these sermons. Moreover, his science is eighteenth-century science; it needs adjustment at a number of places. Still, these sermons preached by the founder of Methodism are profound resources for the ideas discussed in this and other chapters of this book. We commend them to your reading and thoughtful, prayerful consideration.

ENDANGERED SPECIES AND HABITATS

Martin LaBar and Donald D. Wood

And God saw every thing that he had made, and, behold, it was very good.

—Genesis 1:31 KJV

Behold the fowls of the air: for they sow not, neither do they reap, nor gather into barns; yet your heavenly Father feedeth them.

—Matthew 6:26 KJV

We have framed this chapter as a series of questions. This Socratic device should be both a clear and succinct way of approaching some of the many issues involved in surveying the state of species and habitats around the globe today.

What Is an Endangered Species?

Let us first determine what a species is. Some controversy exists among scientists about this concept; some scientists even reject it altogether. However, we will use a definition often agreed upon and used in biology texts: "Species are groups of interbreeding natural populations

> The man who believes things are there only by chance cannot give things a real intrinsic value. But for the Christian, there is an intrinsic value. The value of a thing is not in itself autonomously, but because God made it. It deserves this respect as something which was created by God, as man himself has been created by God.
>
> —Francis A. Schaeffer

that are reproductively isolated from other such groups."[1] In other words, a species is a group of living beings alike enough to interbreed and live in locations such that they are all members of a single population. The International Union for Conservation of Nature (IUCN) estimates between eight million and fourteen million species exist on earth today.[2]

An endangered species is a species at risk of becoming extinct in the near future. The IUCN estimates seventeen thousand or more species are currently endangered.[3]

What Is an Endangered Habitat?

A habitat is a place where organisms live. A suitable habitat requires a specific range of climate or, for underwater habitats, a range of temperature and salinity. For many organisms, a suitable habitat must provide shelter and access to water and nutrients. For many plants, it requires a minimum amount of sunlight. It may require some sort of mechanism for dispersal of the organisms or their reproductive structures.

Biologists agree that most endangered species are endangered because of loss of suitable habitat. Today, habitat loss is usually due to human activity. A forest may be turned into a housing development or wetland may be drained. Sometimes even climate is altered on a significant scale. This, too, can be due to human activity.

An endangered habitat is one that has lost a significant proportion of its area, or is likely to in the near future. Many scientists believe we should concentrate on keeping habitats from becoming endangered rather than species because doing the one will accomplish the other, and because we seldom, if ever, know all the species living within a given habitat. However, the public at large is generally more concerned about preservation of endangered species than of endangered habitats.

Although species endangerment is intimately related to habitat endangerment and the one usually involves the other, the two are not completely identical. Considering a particular species, a competing species may be brought in endangering the first species without loss of habitat for other organisms. Or, a species on which another species depends may become endangered or extinct (due to a new disease, for example) and, in turn, that species becomes endangered or extinct as well, again without a loss of habitat for unaffected organisms.

Here, we will use the term *species*, except when discussing habitat alone. But we ask the reader to remember that the best way to preserve endangered species is to preserve endangered habitats.

What Does the Bible Say about Endangered Species?

Nowhere does the Bible specifically command that species be preserved. It would be a mistake, however, to conclude from this that preserving species from extinction is not part of God's plan and desire for humans in our appointed work of creation care. Because this is treated in the early chapters of this book, we will mention here only a few points briefly as a reminder.

All Is God's and We Are Stewards

Genesis 1:31 reports God's evaluation of the entire creation as "very good." One clear implication is that God views the creation as worth preserving.

God called upon Noah and his family to preserve not just humans but "every kind"—every species that would come to them—from the destruction of the coming flood (Gen. 6:19–20). This included species they would not have considered important but which God did value.

Psalm 24:1 affirms that the earth, along with "everything in it," belongs to God. And Psalm 104, one of the finest nature poems ever written, describes the great diversity of the creation God made "in wisdom" including water organisms "both large and small" (vv. 24–25).

The scouts examined and stated the land to which God was leading Israel was fertile and agriculturally productive (Num. 14:7–8). God gave Israel special instructions to care for the land, even commanding them to let it lie unsown every seventh year (Lev. 25:1–22). Partly because of Israel's disobedience of this specific instruction, they were taken into captivity for seventy years (2 Chron. 36:21) so the land could have the rest God had told them to give it. Thus, God showed concern for the land, the habitat itself.

We Benefit from Responsible Stewardship

We should be careful with non-human species not only because we are stewards of God's creation, but also for other scriptural reasons. One of these is that nature is an excellent teacher. The book of Proverbs refers many times to nature as a guide for humans. A famous example is the list of the diminutive four—ant, coney, locust, and lizard—whose ubiquity and "wise" tenacity belie their individual small size and weakness (Prov. 30:24–28). The wisdom teacher's list of four strutters immediately following, includes three animals and one human—lion, rooster, he-goat, then king (30:29–31).

Jesus also used nature to teach in this way. He used "the birds of the air" and "the lilies of the field" as examples of God's care for all creation. Why, then, should we fret that God may forget us (Matt. 6:25–34)? Jesus' parables of the soils (Luke 8:5–8), the growth of the mustard seed (Mark 4:30–32), and of the weeds (Matt. 13:24–30) are other well-known examples.

This should not surprise us. Psalm 19:1–4 and Romans 1:20 both tell us that observation of nature is one way humans find out about God. If this is true, then destroying part of nature or allowing endangered species to become extinct must interfere to some degree with our ability to hear and see what God wants us to know. For example, if humans saw God's handiwork in the vast flocks of passenger pigeons previously found in eastern North America, could hunting them to extinction have kept some from finding Christ as Savior? It is a question worth asking.

God's Eschatological Goals Include Restoration of Creation

The famous christological hymn of Colossians 1:15–20 says that "in [Christ] all things hold together" (v. 17). It also says he is working to reconcile all things to himself and to make peace through the blood of the cross. As Christians, we are to be ambassadors of this reconciliation (2 Cor. 5:18–20). Should we not also participate in Christ's work of sustaining "all things," including endangered species and biological communities?

In many ways, Scripture tells us God made a good creation, including all types of living things. Under God, we are to do our best to preserve as much as possible of the diversity God has left to our care. This does not mean biological preservation is the highest priority or even that it is possible. Many species have become extinct with no human contact or interaction of any kind. Some species, such as the moa of New Zealand and some North American mammals, were driven to extinction before the humans in those areas were aware of God's revelation through the Bible. Still, we have an obligation to do our best; we do know what the Bible says.

Are There Other Reasons for Preserving Endangered Species?

We may list other reasons for helping preserve the diversity God created. Though presented neither directly or indirectly in Scripture, they do follow from taking seriously the scriptural mandate for stewardship.

Ecology

Many species are integral parts of larger biological communities. If they disappear, other species or even their entire communities will also be put in danger.

Medicine

Many medicines have been discovered in obscure organisms. Taxol, found in the bark of a Pacific yew tree, was discovered in 1967 to be an effective cancer-treating drug. *Forsteronia refracta*, a rare plant from the Amazonian rain forest, produces chemicals showing promise in treating breast cancer. The Plant Conservation Alliance reports that while more than 40 percent of the medicines we use are derived from plants, only about 2 percent of flowering plant species have so far been examined for possible medicinal uses.[4] What discoveries may still lie ahead?

Food

All our food plants were domesticated from wild ones. With continued selection of seed from the best individual plants, modern cultivated plants are often quite different from their ancient ancestors. Some cultivated plants have become less disease-, climate-, or insect-resistant. Breeding or genetically engineering the modern crop plants to re-acquire such resistances by using their wild relatives could be for some people the difference between life and death by starvation. If the wild plants were not preserved, this would not be possible.

Some plants not currently utilized may become food crops in the future. For example, over the last few decades, interest in amaranths as a source of grain has been increasing. In livestock, the bison again is becoming a source of meat in the American Midwest.

Economic Assets

Some endangered species and the habitats in which they live are tourist attractions and bring significant income. The national park system in the U.S. has helped preserve bison and elk for human enjoyment and, in the process, has provided income for communities near the parks. For a number of African nations, the same is true of the national park habitats of lions, rhinoceroses, zebras, and many other animals.

Aesthetics

Many plants and animals are beautiful to see, hear, or even smell. Think of a herd of elk or bison in North America, or wildebeest on the African Serengeti, songbirds or orchids in a tropical rainforest, or seals, dolphins, and whales off the coasts of North America. Some beautiful sites actually are habitats; a coral reef community is a prime example. Do we have the right to deprive future generations of such experiences by our neglect or unwise actions?

This relates, of course, to God's revelation through nature. But we humans usually try to preserve artifacts for their aesthetic and historical values, independent of their usefulness in revealing God. One example is the international outcry when the Taliban destroyed ancient Buddhist statues in Afghanistan. Should not natural beauty also be preserved?

Ethics

Do animals have a right to exist? To argue they do may be controversial, but it is a difficult argument to resist in the case of some species that appear to be highly intelligent, communicate with each other, and perhaps even have a moral sense—species such as elephants, great apes, and cetaceans (dolphins and whales). Could the loss of one or more of these endangered species even mean the loss of potential ethical understanding? We have no way of knowing.

We are not suggesting that any non-human animal should be given human rights. The Bible is clear that humans are especially in God's image. God came to earth as a human, not as some other creature. We bear God's image in a unique way and the responsibilities that go with that. But, just as an artist creates paintings, concertos, or poems that reflect her own personality, so much so that experts can identify them as her creations even without a signature, in the same way, surely, God leaves his stamp, some part or aspect of his image, in non-human living things, even in the very rocks and soil of the earth itself.

Does the Imminence of Christ's Return Remove Our Obligation to Preserve Endangered Species?

Our first response to this question is we cannot be sure Christ will return in what we would measure as the near future. Christians have been saying he would return soon almost since the day of his ascension, and so far all have been wrong. His return may be today, or it may be centuries or millennia in the future. However, even if it is today, Christ still expects to find us being good stewards of the natural world.

To invoke Christ's return in such an argument really is to say that his return should motivate us toward evangelism at the expense of all other activity. His return should motivate us to evangelize, but even if we had no such promise, we ought still to be trying to win others because each of us is going to die. However, we need to be engaged in helping the poor, in making sure our relationships with others are Christlike, in study, in worship, and in many other actions as well, because they are also part of our witness to the gospel, part of our evangelism. Even if we knew Jesus' coming were tomorrow, this would not relieve us of all duties other than direct evangelistic witness in words. If we really have a duty to preserve endangered species, then it, too, remains independent of the timing of Christ's return. We believe we have such a job.

Should We Try to Preserve Even Pests and Weeds?

There is no biblical mandate for completely exterminating thistles, for example, but neither does the Bible prohibit us from working to remove them from our fields. The Old Testament has a number of admonitions to work hard, most colorfully among the pithy sayings of Proverbs. For farmers, part of this involves doing as much as they can to keep weeds out of their crops. In Christ's story about the wheat and the weeds (Matt. 13:24–30), the servants were not expecting to find weeds growing among the wheat. It may not have been possible to weed a wheat field by hand, but steps were taken to keep weed seeds out of the wheat seed for the next season.

When the Israelites crossed the Jordan, Moses told them God would not drive out the nations living there all at once, so the dangerous animals would not increase, as they would have done with no humans living there (Deut. 7:22). This indicates that, as in the case of weeds, it is legitimate to control the populations of dangerous animals (large predators and poisonous reptiles, for example) living close to humans.

From these and many other scriptural passages, we learn that part of good stewardship in this world held hostage to our disobedience is to keep weeds out of our crops and dangerous animals away from our children. We are expected to do this. We believe this principle—regard for human life above those organisms that would harm us—also applies in other situations. Clean water, safe disposal of sewage and other wastes, elimination of infectious diseases—all require destruction of microorganisms. But when not controlled, these kill many thousands of humans; controlling them is part of our stewardship mandate.

One threat to endangered species (though not immediately to humans) is non-native, invasive species, usually introduced by deliberate or inadvertent human action. Examples are the zebra mussel and quagga mussel, both introduced accidentally to North America from their origins in Eastern Europe. They out-compete native mussel species, threatening their survival.

Some invaders threaten not just one species but an entire habitat. One example is the crown-of-thorns starfish which attacks corals. Coral colonies form reefs, creating a habitat for many other species. When the starfish destroy the coral, these lose their habitat too. An example from the plant kingdom is the overrunning of many locales in the southeastern U.S. by kudzu, imported from eastern Asia. If our call to stewardship includes protecting endangered species, then it includes trying to prevent invasive species from driving native organisms to extinction.

In some situations, predators can threaten the extinction of some species from a given area. Of course, we cannot (and should not try to) prevent all predation. However, if a predator seems likely to exterminate a species entirely, we should consider intervention. One current example is the multiplication of feral cats in rural Wisconsin and their severe impact on the numbers of several songbird species.

We believe it is appropriate to try to keep weeds, pests, diseases, and other threats to humans, crops, and domestic animals at bay. We also should try to rectify our mistakes in introducing invasive species where they have no natural controls, and where they threaten to overwhelm native species. This will not drive them to extinction; it will merely remove them from inappropriate locations.

What if Human Needs Conflict with Preserving Endangered Species or Habitats?

Several African populations now do not have enough land to sustain themselves with traditional agriculture. Should they be allowed to grow crops in national parks adjacent to where they live? Should they be allowed to hunt for food in these national parks? These are tough questions. One of the purposes of national parks is to preserve endangered species and habitats. On the other hand, who are we in North America to tell Africans, living very simply by comparison, they cannot use available land to feed themselves? It may be possible to change the way people facing this dilemma make their living, so they do not endanger their national parks and the animals living there but such decisions are for Africans to make in Africa.

What about North America? Should we limit our housing and road development for the sake of the land? Do we have the right to impose a simplified community where many living things are not welcome around our houses in the form of lawns? Do we have the right to deplete our water tables for the sake of building great cities in our deserts?

One reason such dilemmas arise is the recent rapid increase in human population, giving world-wide impetus to habitat destruction. Some Christians believe that God's command to be fruitful and multiply (Gen. 1:28) means human population growth should not be checked. We believe this command is one of the few we have consistently kept and that it does have limits. Responsible stewardship in all areas includes sustainable levels of human population that do not endanger the earth and its ecosystems. If, by our numbers, we seriously degrade the earth we, too, will lose.

Is Global Warming a Threat to Species and Habitats?

We understand the controversy on this issue has increased recently rather than abated, but we believe global warming is a threat. As with all threats, it is possible to exaggerate it and some have exaggerated recent climate change. It is also true that past climate changes have occurred without human intervention. Still, many in the scientific community are convinced the earth is currently getting warmer and much or all of this current warming is due to human activity.[5] These scientists identify the release into the atmosphere of several substances that slow the escape of heat from the earth's surface as the main culprit. Any resulting significant temperature change is likely to affect the earth's climate in a number of ways. Such changes can be expected not only to eliminate some species, but also change, eliminate, or relocate entire habitats.

The main objections to dealing with climate change, besides not believing in it, are resistance to change in economic and other areas. One example is burning petroleum for fuel which adds to the greenhouse gases in the atmosphere. Therefore, we believe North American society as a whole needs to rethink our dependence on the gasoline-powered automobile as our preferred means of transportation. Will the change to mass transit and other types of personal vehicles be simple? Of course not, but adopting gasoline-powered automobiles on a large scale in the first place was not easy, either. We hope this discussion among Christians can be continued in Christian love and in the recognition that global warming or not, God expects us to use all the earth's resources carefully and wisely.

Is Preserving Endangered Species the Christian's Highest Priority?

Surely not. The New Testament says a great deal more about evangelism, prayer, and other subjects than it says about other species. The Bible as a whole includes our stewardship mandate together with many other responsibilities. From this, we conclude that this duty should not take precedence over commands of Christ and other responsibilities.

However, we believe our stewardship mandate includes doing what we can to preserve endangered species. We will not be able to preserve all species, but that is not an excuse to do nothing. We have not yet done as much as we should; therefore, we should be doing more. As Oswald Chambers put it: "If we are children of God, we have a tremendous treasure in nature and will realize that it is holy and sacred. We will see God reaching out to us in every wind that blows, every sunrise and sunset, every cloud in the sky, every flower that blooms, every leaf that fades."[6]

Suggestions for Reflection and Action

1. Study the Scripture passages listed in the section, "What Does the Bible Say about Endangered Species?" (pages 171–173). Study to see whether you agree these passages indicate we are under God's mandate to try to preserve endangered species. If so, why? If not, why not? This could be a project over several weeks, working with others in a small group.

2. Begin or continue actions often suggested to cut consumption and slow habitat destruction: (1) participate in local recycling programs, (2) use cloth or other reusable shopping bags, rather than plastic ones, (3) combine daily errands into one outing and plan the shortest route beforehand, (4) walk wherever possible—it's also excellent exercise, and (5) do all these as much as possible with friends and neighbors.

3. Learn and practice ecologically friendly ways to grow gardens and decorative plantings. Learn how to make your landscaping a better habitat for small creatures. Keep yourself regularly updated on positive ecological information and action.

For Further Reading

Bratton, Susan Power. *Six Billion and More: Human Population Regulation & Christian Ethics*. Louisville: John Knox Press, 1992.

The title suggests the thrust of the book. Reading it will open your eyes to a diversity of issues and Christian viewpoints on the implications of human population increase.

Rolston, Holmes, III. *Conserving Natural Value*. New York: Columbia University Press, 1994.

Rolston, a Christian and one of the founders of the field of ecological ethics, discusses many of the ways nature is valuable to us. Though not written exclusively for a Christian audience, this book empowers a new appreciation for the "very good" of God's creation.

Schaeffer, Francis A. and Udo W. Middlemann. *Pollution and the Death of Man*. Wheaton, Ill.: Crossway Books, 1992.

Schaeffer argues there really is an environmental crisis and that Christians should do something about it. He also points out wrong solutions, including pantheism.

Wilkinson, Loren. *Earthkeeping in the Nineties: Stewardship of Creation*. Eugene, Ore.: Wipf and Stock, 2003.

Wilkinson reviews the history and theology of Christian stewardship then treats inadequate approaches to stewardship, such as ecofeminism and new-age movements. He also includes suggestions for practical, effective action.

Notes

1. Ernst Mayr, *Populations, Species, and Evolution: An Abridgment of Animal Species and Evolution* (Cambridge, Mass: Belknap Press, 1970), 12.

2. International Union for Conservation of Nature, State of the World's Species, http://cmsdata.iucn.org/downloads/state_of_the_world_s_species_factsheet_en.pdf (accessed August 11, 2009).

3. Ibid.

4. Lissa Fox and Peggy Olwell, "A Living Gold Mine," (Washington, D.C.: Plant Conservation Alliance, 2005), http://www.nps.gov/plants/lgm.htm (accessed March 16, 2010).

5. Intergovernmental Panel on Climate Change, "Climate Change 2007: The Physical Science Basis," http://www.foxnews.com/projects/pdf/SPM2feb07.pdf (accessed January 26, 2010).

6. Oswald Chambers, *My Utmost for His Highest: An Updated Edition in Today's Language* (Grand Rapids, Mich.: Discover House, 1992), entry for February 10.

A CALL TO ACTION

Matthew and Nancy Sleeth

Why do you see the speck in your neighbor's eye, but do not notice the log in your own eye? Or how can you say to your neighbor, "Let me take the speck out of your eye," while the log is in your own eye? You hypocrite, first take the log out of your own eye, and then you will see clearly to take the speck out of your neighbor's eye.

—Matthew 7:3–5 NRSV

We were environmentalists before we were Christians, but we did not act like Christians until faith inspired us to scale back our lifestyle. How did our faith and environmental journey begin? Like most of the physicians Matthew worked with a decade ago, he did not believe in God. He thought science and rationality were the keys to unlocking happiness. We didn't go to church, own a Bible, or hang out with people who did. Matthew was raised a Protestant; Nancy was raised in a Jewish home. Reactions from

> If the Creator and Father of every living thing is rich in mercy towards all . . . how is it that misery of all kinds overspreads the face of the earth?
>
> —John Wesley

family and friends convinced us religion was only a means for people to justify their own prejudices. It was of no use to us.

For the twenty years following our wedding, we pursued the American dream. Matthew enrolled in college; ten years later he was a practicing physician. We moved to a postcard-perfect neighborhood on the coast of New England. Matthew was doing something he loved, and he relished the prestige, the paycheck, and the respect attached to being director of an emergency room and chief of staff of the hospital.

Our Teshubah: Our Turn-Around

One February, our family vacationed on an island off the coast of Florida. Warm, beautiful, and quiet, it had no cars and no streetlights; its narrow roads were sand. Once the children were tucked into bed, we would sit on a balcony facing the water, the stars shining brightly, palm trees rustling in the lovely Gulf breeze.

In that tranquil setting, Nancy asked a question that was to change our lives forever. "What," she wanted to know, "is the biggest problem facing the world today?" After thinking for a minute, Matthew answered, "The world is dying."

Matthew gave this answer not because he was a biologist or ecologist, but simply based on the observations of his life. There are no more elm trees on Elm Street, no more chestnuts on Chestnut Lane, no caribou in Caribou, Maine, and no more blue pike in the Great Lakes.

These are not rare or exotic species that have fallen prey to extinction. A few decades ago, the blue pike was the most numerous and commercially harvested freshwater fish in the world. The fields and fence rows in which many of us played as children have been bulldozed and planted with houses. When we tried to find the ford where Matthew had proposed, we could not find even the stream. It had been buried under a subdivision.

Similar changes have occurred in humans. We get more cancers, more asthma, and more autoimmune diseases. In response, we build bigger hospitals and develop more medicines and more radiation treatments. Healthcare becomes more expensive every year, yet we do not ask, "Is it because our environment is making us sick?"

We sat in the tropical quiet for awhile, then Nancy asked a much more difficult question, "What are we going to do about it?" Matthew did not have an immediate answer. How does one get one's mind around the dying of a planet?

About this time, within a single week, three women were admitted to Matthew's emergency room. All three had breast cancer; all three were in their thirties; all three died. One died with seizures right there in the ER, and Matthew had to tell her young husband and two small children she was gone. Not long before that, Nancy's brother had drowned. Then a patient Matthew had treated several times began to stalk our family. We had all the things that were supposed to make us happy, but our life felt barren. We had no spiritual compass to direct us through murky waters.

We live in a world of measurements, yet the evil and pain we were witnessing—like love, hope, and even God—could not be measured in a double-blind study. All these defy the quantification of modern science; we had to look in new places. We began reading through various sacred texts: the Ramayana, the Bhagavad-Gita, the Koran. These contain many truths, but we did not find the answer to the question, "How do we save a dying world?"

One afternoon, Matthew walked into a hospital patient lounge and sat on a couch next to a coffee table. On one corner, he saw an orange Bible placed by the Gideons. He took it home, read it, and found the truth he had been seeking. He became a believer—a follower of Jesus. Before, he had always assumed that science or business or government would provide the answers. Reading the Bible, he realized his heart must change before he could make other significant changes—changes that would require sacrifices. This book required him to look in the mirror. Matthew 7:3–5 said we were not to worry about the speck in our neighbor's eye until we had removed the plank from our own eye. Jesus, whom we now sought as our guide, told us to be meek, humble, compassionate, thankful, forgiving, and "above all, clothe[d] . . . with love" (Col 3:14 NRSV).

Wanting to see whether the Bible had anything to say about caring for the earth, Matthew read it from cover to cover, underlining everything that had to do with the environment, stewardship, and caring for creation. He ended up with an underlined Bible, from Genesis to Revelation. Clearly, we needed to scale back our lifestyle, focus less on getting and more on giving, and consume less so we could serve more.

Eventually Matthew got back to Nancy with an answer to her second question, the one about our response to the fact that our earth is dying. He told her he would quit his job and start working full time in a job description that did not exist. Would he be a green doctor? A creation care minister? An eco-evangelist? We didn't even know how to describe Matthew's new calling, but we felt certain the call was from God. Though we were leaving the safety of a paycheck, Matthew was not leaving health care. He was simply shifting to health care on a global scale, trying to help avert the biggest health care crisis our planet has ever witnessed.

The home, the car, the job—they're all gone. Our family moved to a house the size of our old garage. Don't feel sorry for us; we had a doctor-sized garage. We reduced our electricity usage, fossil fuel consumption, and trash production to a small fraction of the national average. These changes were motivated by faith. Reading through the Bible, we found eternal answers to today's problems. In its pages, we found that God not only loves us; God also loves the tree outside our window, and all the birds, squirrels, and insects that dwell in its branches.

Jesus' Answer: The Good Samaritan

When we speak at churches and colleges, we often retell the parable of the good Samaritan to illustrate how Jesus wants us to act when confronted with twenty-first-century issues such as Earth stewardship.

The parable begins with Jesus being asked a very important question that all of us should consider: "Teacher, what shall I do to have eternal life?" (We will paraphrase this conversation, but only slightly. See Luke 10:25–37 for the whole story.)

Jesus answered the scholar with a question of his own, "What does the law say?"

The scholar replied, "Love the Lord your God with all your heart, with all your soul, with all your strength, and with all your mind— and your neighbor as yourself."

Jesus commended the man: "Good answer! If you do this, you will live."

Jesus here affirmed the code for living found in the Torah (the Old Testament law). We summarize it in fifteen words: (1) love God with all your heart, soul, strength, and mind; (2) love your neighbor as yourself.

However, like many of us, the scholar was looking for a loophole and asked, "Who is my neighbor?" Jesus then told the parable of the good Samaritan to define for us our neighbor. In so doing, Jesus taught that it is not the letter but the spirit of the law that is important to God.

As Jesus began the story, he told of a man traveling from Jerusalem to Jericho, a common journey on a road many of his hearers would have traveled themselves. This man was mugged, stripped, and left for dead. A priest from his own faith, one of his countrymen, came along. Rather than stopping to help the wounded man, though, he passed by on the other side of the road. Not long after, another member of the temple professional staff, a Levite, came along. He saw the helpless man, and his heart was moved by compassion. He thought, *The Romans should patrol these roads better. The government should install better street lighting. When I get back home, I'm going to raise awareness and blog about this.* But of course, such pious concern did not do the injured man any good. As each of them continued on his own journey, the victim still lay moaning alongside the road.

Finally, a third man came along. This traveler was not of the mugged man's faith. He was a Samaritan, and in the first century, the Jewish and the Samaritan people despised each other. Now, the Samaritan was not walking; he rode a donkey. If given the choice between walking and riding on a long journey with a significant elevation drop through dangerous territory, most of us, like the Samaritan, would choose to ride.

The Samaritan had more resources—he was riding rather than walking—so it is even more striking that he stopped to help. Seeing the wounded man, he dismounted his donkey, anointed the injured man with oil and bound up his wounds, gave him wine to ease the pain. He then lifted him upon his own donkey, and walking beside him, supported him until they reached the nearest inn. Finally, the Samaritan not only paid for the man's immediate care, but left an advance, even promising to return and cover any and all expenses until the man he had rescued was fully healed.

It is worth pausing to note here that, as an ER physician, Matthew saw some thirty thousand patients. Only once did someone come in and pay the hospital bill for a stranger.

Having finished this astonishing story, Jesus now asked the scholar, "Which of these three travelers acted as 'neighbor' to the man who was mugged?" He could not bring himself even to say the word *Samaritan*, so he responded, "He who showed mercy." Jesus' command, then, is for us as it was for his questioner, "Go and do likewise!"

Putting Our Faith into Action

What do the actions of the Good Samaritan teach us about how we should approach environmental problems today? In this parable, Jesus is showing us a continuum of compassion: how each of us can respond to problems, such as the environment. Implied in the story's sequence is that the first man, the priest, was like those of us who deny a problem even exists. We close our eyes and don't even stop to consider the consequences of our actions.

The second traveler, Jesus intimated, was where most of us are: He saw that a problem existed, but did not want to take any action that would have involved a degree of personal sacrifice. Like the Levite, many of us see the hardship caused by environmental problems, particularly for the poorest among us. Our hearts even may be moved to compassion, but we do little if anything to help because we don't want to be inconvenienced.

The Samaritan is the example Jesus wants us to follow. Only the Samaritan, the man thought least likely of these three to view the fallen Jewish man as his neighbor, took action. He saw the need and had mercy. Jesus tells us, as he told his questioner, to act with compassion.

To have any lasting effect, the compassion of our hearts, too, must be moved to action on behalf of our neighbors near and far and on behalf of all God's creation. We may find it dangerous. We may have to use our own resources. It may be inconvenient. It may be expensive. We may be ridiculed. We may have to take ongoing responsibility and make personal sacrifices. But such is the path to eternal life.

Everyone is our neighbor, including foreigners, strangers, people who hate us, and future generations. Perhaps the most important lesson of the Good Samaritan—the action that can separate us from the priest and the Levite—is that we must get off our donkey before we can become part of the solution. The future will not be saved by our good intentions. It will be made better or worse only by our actions.

We show our love for God by loving our neighbors. Now, in our family, every time we buy anything or take any action, we ask two questions: Will this help me love God? And will this help me love my neighbor? The answers always lead us to right action.

If we take shorter showers, car pool, or plant a tree, no one necessarily notices or thanks us. But if we do these things as acts of service to God and of protection for our neighbors, then we grow as loving, spiritual beings. Love is the great hope the church offers the environmental movement.

Our nineteen-year-old daughter, Emma, recently served four months providing palliative and hospice care to the poor in southern India. Average annual income in India is six hundred twenty dollars. Only four out of one hundred people have access to telephones, and thousands of children scavenge for food in trash heaps. Compared to the dying cancer patients Emma visited—people who cannot buy their way out of the effects of environmental degradation—we have had to ask ourselves again, "How thoughtful are we? What kind of neighbors are we?"

The Benchmark

In Jesus' parable, the Samaritan compared himself to the wounded man, applied the Golden Rule, and was compelled to act. Jesus relentlessly told (and still tells) his listeners to observe the plight of less fortunate neighbors and take steps to help. Christ asks us to change our behavior. To do that, we need a benchmark to know what our behavior is. Will you take courage now to ask yourself, to answer honestly, and to record your answer, "What is my behavior toward God's creation?"

For the majority of us, our relationship to the created world is not one of caretaker or steward. Our typical reaction to nature is to not see it, or to see it only when a vacation or a sporting activity takes us into contact with it. It is sobering for [us] to admit that [we] can identify more [makes] of automobiles than [species of] trees.[1]

To move from just thinking about the care of creation to actually doing something about it, we have to become aware of what kind of action is needed. That may begin with something as simple as thanksgiving.

When we are truly grateful, we give God thanks for our blessings. When we are ungrateful or feel a sense of entitlement toward material blessings, we tend to ignore or not give thanks. Many of us give thanks for our food. . . . Few of us recognize, however, that people work, fight, and die to bring us energy.

Energy—electricity, wood, coal, gasoline, propane, and oil—is like food. It is a blessing, and it sustains us. . . .

When was the last time you bowed your head in thanks when filling your car with gasoline? If you haven't done so, is it because you don't think it is a blessing? Do you feel entitled to fill up?[2]

We believe humanity stands at a great crossroads. We hold the fate of God's creation in our hands. This is not because there is no God, or because God is not all-powerful, does not love, or is not in control. Rather, it is the result of God having made us in his own image with the freedom to choose. We are free to choose life or death, light or darkness, and the very fate of our own souls. With this awesome responsibility comes the stewardship, not only of the natural world we inhabit, but also of the fate of our children and our children's children.

The preceding chapters in this book encourage us to become better stewards of the gifts on loan to all of us. The less we fill our homes with things, the more content, joyous, and spirit-filled our lives grow. As our conclusion to this chapter, you will find a list of specific actions you can take today, this week, this month, and this year—actions that can help you move from passive understanding and compassion to sustained stewardship involvement.

Together with Jesus' parable of the good Samaritan, Matthew 7 has inspired our journey. There, Jesus said, "Ask, and it will be given you; search, and you will find; knock, and the door will be opened for you. For everyone who asks receives, and everyone who searches finds, and for everyone who knocks, the door will be opened" (Matt. 7:7–8 NRSV). May these two important passages encourage you to begin, or to continue, the journey with us.

Putting Your Faith into Action

First, pray this prayer, or a prayer in your own words; as you pray, be sure you mean the words you are praying:

Dear Heavenly Father, Creator, and Sustainer, give me the knowledge and will to honor you by using resources wisely in my home. Help me to preserve rather than destroy; teach me to conserve rather than waste. Remind me that everything I possess is on loan from you. Help me to create a God-centered home that I share freely with others. Strengthen my desire to become a better steward of your abundant blessings.

Second, take account of how much electricity, fossil fuels, and trash your family consumes, then set goals for reduction. If you can do 10 percent better each year, you are on the right path.

Third, prayerfully and carefully consider the following possibilities for your own further action, individually or with others in your family, your church, and your communities. Add to this list, also, as God continues to bring to your attention details of the physical environment in which you live, and people among whom you live, individuals and groups with whom you have the opportunity to act as the Samaritan did in Jesus' parable.

Lord, Help Me Today To:

- Reduce my shower time by two minutes.
- Turn off the faucet while brushing my teeth and shaving.
- Turn my water heater down to a lower setting.
- Turn my thermostat up three degrees (in summer) or down three degrees (in winter).
- Turn off the lights, TV, radio, and stereo when I leave the room.
- Use cold water if I use the garbage disposal.
- Turn my refrigerator and freezer to a warmer setting.
- Run only full loads in the dishwasher.

Lord, Help Me This Week To:

- Read Psalms 23, 24, 104, 147, and 148.
- Find out if my public utility company offers a green power option and sign up.
- Avoid using aluminum foil and plastic wrap.
- Change at least five light bulbs in my home to CFL (compact fluorescent light) bulbs.
- Wash my clothes in the coolest water possible and only run full loads.
- Donate a box of books to the library.

- Buy only tree-free toilet paper, paper towels, and tissues made from recycled paper.
- Air-dry my laundry—if I use the dryer, use the moisture sensor option.
- Pre-cycle by buying minimally packaged goods and choosing reusable over disposable.
- Cut back on the amount of junk mail I receive by registering at https://www.dmachoice.org.

Lord, Help Me This Month To:

- Stock up on handkerchiefs, cloth shopping bags, and cloth napkins so I can kick the paper habit.
- Clean out my closets and donate clothes I have not worn in the past year.
- Install low-flow showerheads.
- Switch to environmentally friendly cleaning products.
- Clean or replace air filters throughout my house.
- Wrap my water heater in an insulating jacket if it is hot to the touch or more than ten years old.
- Caulk and weather-strip around my windows and doors to plug air leaks.
- Unplug the TV, computer, and audio equipment when not in use, or put them on a switch-controlled power strip.

Lord, Help Me This Year To:

- Donate my old cell phone, computer, or printer to a good cause.
- Make or purchase insulated window treatments.
- Purchase only the most efficient Energy Star items when appliances and lighting fixtures need to be replaced.
- Ask my utility company to conduct an energy audit on my home and follow up on their advice.

- Insulate my walls and ceilings to save up to 25 percent on my energy bill.
- Use the money I save to advance your kingdom.

Notes

1. J. Matthew Sleeth, *Serve God, Save the Planet: A Christian Call to Action* (Grand Rapids, Mich.: Zondervan, 2007), 60.

2. Ibid., 60–61.

ABOUT THE AUTHORS

Christina T. Accornero is associate professor of religion and chair of the division of religion at Southern Wesleyan University (SWU), where she has taught since 2008. She earned a Ph.D. in intercultural studies from Fuller Theological Seminary in 1998.

Christopher T. Bounds is associate professor of theology at Indiana Wesleyan University (IWU), where he has taught since 2002. He earned a Ph.D. in systematic and Wesleyan theology from Drew University in 1997.

Joseph Coleson is professor of Old Testament at Nazarene Theological Seminary, where he has taught since 1995. He earned a Ph.D. in Near Eastern and Judaic studies from Brandeis University in 1982.

Ronald R. Crawford is science supervisor at Grove City Christian School, Grove City, Ohio, where he teaches upper-level and advanced placement biology and chemistry; he taught previously at IWU, Ball State University, and Oklahoma Wesleyan University (OWU). He earned an Ed.D. in science education from Ball State University in 1990.

Richard L. Daake is professor of chemistry and chair of the department of science and mathematics at OWU, where he has taught since 1976. He earned a Ph.D. in inorganic chemistry from Iowa State University in 1976.

Kenneth D. Dill is associate vice president for spiritual life and university chaplain at SWU, where he has taught since 1990. He earned an M.Div. from Emory University in 1984.

Kelvin G. Friebel is associate professor of Old Testament at Houghton College, where he has taught since 2006; previously, he taught at Canadian Theological Seminary, Regina, Saskatchewan. He earned a Ph.D. in Hebrew and Semitic studies from the University of Wisconsin in 1989.

Kenneth F. Gavel is professor of biblical studies and theology, and chair of the division of biblical/theological studies at Bethany Bible College, Sussex, New Brunswick, where he has taught since 2000. He earned a Ph.D. in systematic theology from the University of Edinburgh in 2003.

D. Darek Jarmola is professor of historical theology and director of the persecuted church ministries program at OWU, where he has taught since 1994. He earned a Ph.D. in historical theology/reformation studies from Southern Baptist Theological Seminary in 1990.

Martin LaBar is professor of science emeritus at SWU, where he began teaching in 1964. He earned a Ph.D. in genetics and zoology from the University of Wisconsin in 1964.

Stephen J. Lennox is professor of Bible at IWU, where he has taught since 1993; he also has served as division chair and as dean of the chapel. He earned a Ph.D. in biblical studies from Drew University in 1992.

Jo Anne Lyon was elected a General Superintendent of The Wesleyan Church in June, 2008. She earned an M.A. in counseling psychology from the University of Missouri-Kansas City, and has done further work in historical theology at Saint Louis University. She has been awarded a number of honorary doctoral degrees. Dr. Lyon is known internationally as the founder of World Hope International, a relief and development agency.

Travis H. Nation is associate professor of biology at SWU, where he has taught since 2002. He earned a Ph.D. in zoology/ecology from Clemson University in 2005.

Susan Rouse is associate professor of biology at SWU, where she has taught since 2005. She earned a Ph.D. in neuroscience from Emory University in 1991.

Matthew and Nancy Sleeth are founders of Blessed Earth, an educational non-profit organization dedicated to the promotion of creation care from a Christian faith perspective. Matthew's degrees in medicine and Nancy's in journalism serve them well as they educate Christians and others about God's creation mandate to serve and guard the earth.

Burton Webb is professor of biology and associate dean of the school of the physical and applied sciences at IWU where he has taught since 1994. He earned a Ph.D. in microbiology and immunology from the Indiana University School of Medicine in 1995.

Donald D. Wood is professor of religion at SWU, where he has taught since 1978. He earned a Th.D. from Fuller Theological Seminary in 1974.

Other titles in the Wesleyan Theological Perspectives Series!

Edited by Joseph Coleson

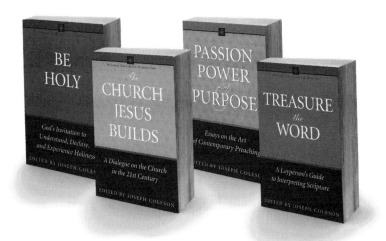

Be Holy
*God's Invitation to
Understand, Declare,
and Experience Holiness*
BKBR18 9780898273724

Passion, Power, and Purpose
*Essays on the Art of
Contemporary Preaching*
BKBK48 9780898273366

The Church Jesus Builds
*A Dialogue on the
Church in the 21st Century*
BKBP36 9780898273496

Treasure the Word
*A Layperson's Guide to
Interpreting Scripture*
BKBT94 9780898274127